Autodesk® Inventor® 2014

A Tutorial Introduction

Scott Hansen

ISBN: 978-1-58503-821-3

SDC Publications

Schroff Development Corporation

www.SDCpublications.com

Schroff Development Corporation
P.O. Box 1334
Mission KS 66222
(913) 262-2664
www.SDCpublications.com

Publisher: Stephen Schroff

Examination Copies:

Electronic Files:

Table of Contents

Index

Notes:

CHAPTER 1

Getting Started

Objectives:

1. Create a simple sketch using the Sketch Panel
2. Dimension a sketch using the Dimension command
3. Extrude a sketch in the Model/Part Features Panel using the Extrude command
4. Create a hole in the Model/Part Features Panel using the Extrude command
5. Create a fillet in the Model/Part Features Panel using the Fillet command
6. Create a counter bore in the Model/Part Features Panel using the Hole command

Chapter 1 includes instruction on how to design the part shown.

1.	Start Inventor by moving the cursor to the 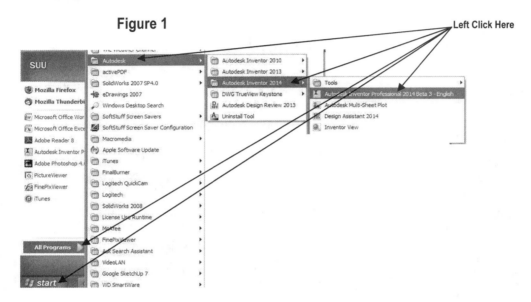 button in the lower left corner of the screen. Click the left mouse button once.

2.	A pop up menu of the programs that are installed on the computer will appear. Scroll through the list of programs until you find Autodesk Inventor Professional 2014.

3.	Move the cursor over **Autodesk Inventor Professional 2014** and left click once.

Figure 1   Left Click Here

4.	Autodesk Inventor Professional 2014 will open (load up and begin running).

5.	The Autodesk Product banner will appear. Left click on the close icon as shown in Figure 2.

Figure 2   Left Click Here

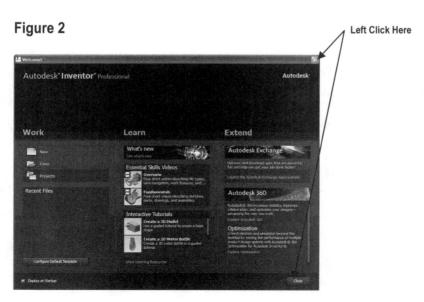

6. Left click on the "New" icon (white piece of paper) located in the upper left portion of the screen as shown in Figure 3.

Figure 3

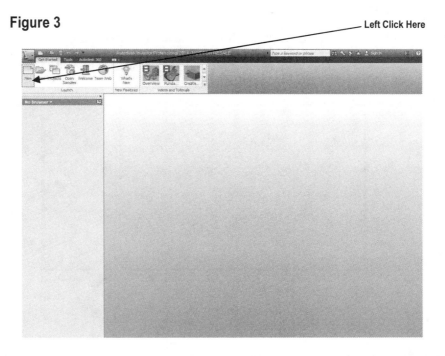

7. The Create New File dialog box will appear. Left click on the **English** folder at the upper left corner of the dialog box. After left clicking on the **English** folder, left click on **Standard (in).ipt** as shown in Figure 4.

Figure 4

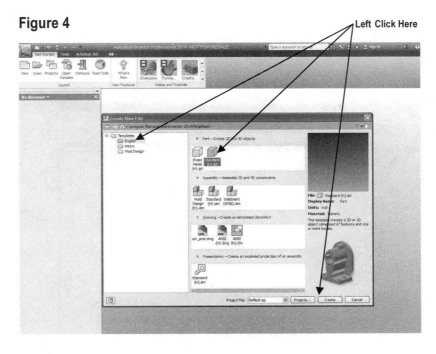

8. Left click on **Create**.

9. Inventor is now ready for use. The screen should look similar to Figure 5.

Figure 5

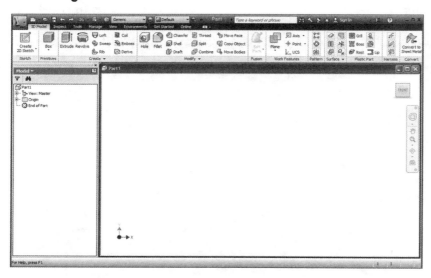

Create a simple sketch using the Sketch Panel

10. Begin a drawing by first constructing a "sketch". Move the cursor to the upper left portion of the screen and left click on the **3D Model** tab (if not already selected). Left click on **Create 2D Sketch**. Select the Front Plane as shown. Move the cursor to the upper left portion of the screen and left click on **Line** as shown in Figure 6. To know what any icon or command will do, move the cursor over the icon or command and wait a few seconds. A yellow banner will appear describing the icons or commands function.

Figure 6   Left Click Here

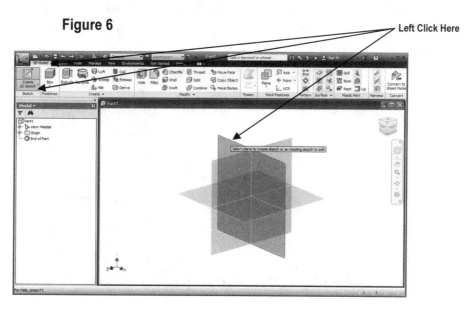

11.   Your screen should look similar to Figure 7. Left click on the **Sketch** tab if needed. Move
        the cursor to the upper left portion of the screen and left click on **Line** as shown in Figure 7.

Figure 7   Left Click Here

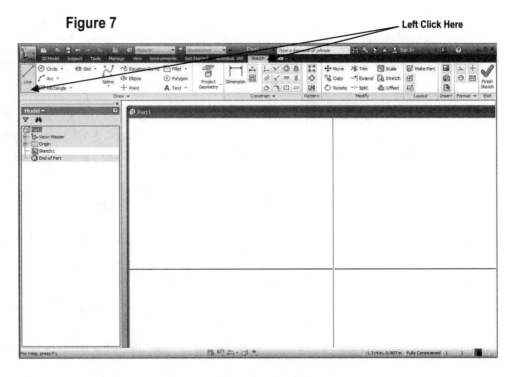

12.   Move the cursor somewhere in the lower left portion of the screen and left click once. This
        will be the beginning end point of a line as shown in Figure 8.

Figure 8   Left Click Here

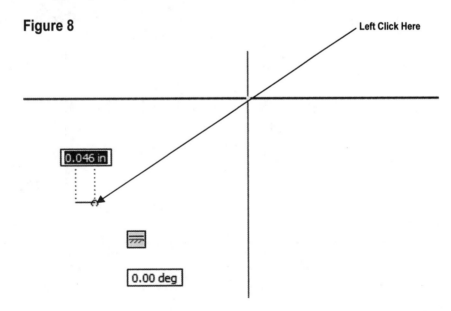

13. Move the cursor towards the lower right portion of the screen and left click once as shown in Figure 9.

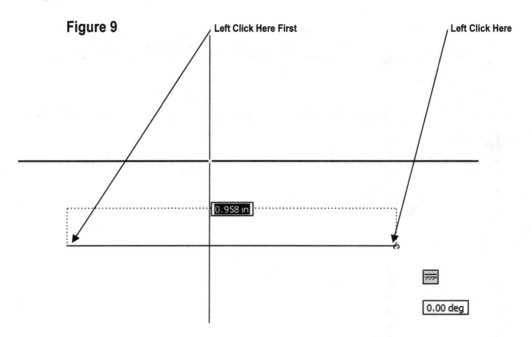

Figure 9

Left Click Here First

Left Click Here

0.958 in

0.00 deg

14. While the line is still attached to the cursor, move the cursor towards the top of the screen and left click once as shown in Figure 10.

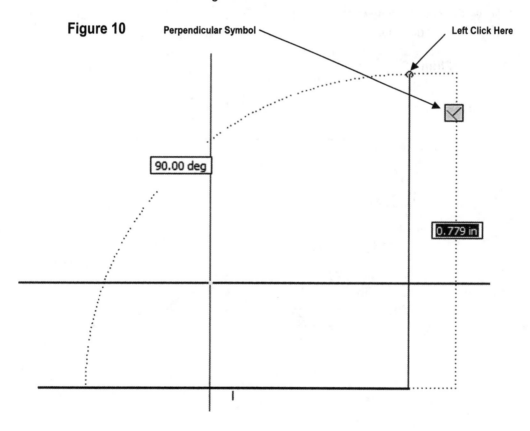

Figure 10

Perpendicular Symbol

Left Click Here

90.00 deg

0.779 in

15. This signifies that the vertical line is exactly 90 degrees (perpendicular) to the horizontal line.

16. With the line still attached to the cursor, move the cursor towards the left side of the screen as shown in Figure 11.

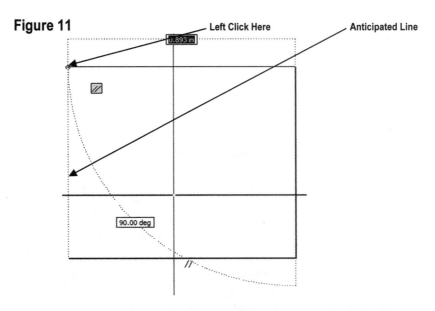

Figure 11

17. Notice the line of small dots connecting the first and last points together. Left click once when the small dots appear as shown in Figure 11.

18. This will form a 90 degree box. Move the cursor down towards the original starting point. Ensure that a green dot appears (as shown in Figure 12) at the intersection of the two lines. This indicates that Inventor has "snapped" to the intersection of the lines. After the green dot appears, left click once as shown in Figure 12.

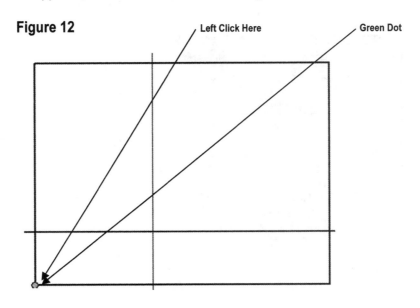

Figure 12

19. Your screen should look similar to Figure 13.

Figure 13

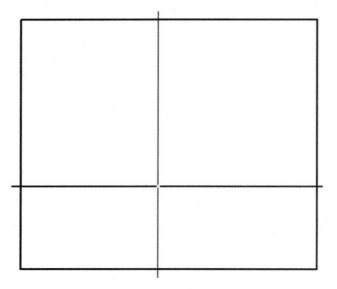

Dimension a sketch using the General Dimension command

20. Right click anywhere around the sketch. A pop up menu will appear. Left click on **OK** as shown in Figure 14. Hitting the **ESC** key will also "get out" of the Line command.

Figure 14

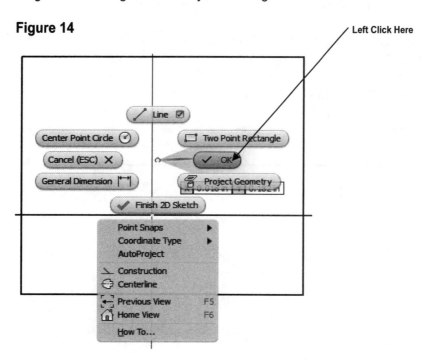

21.   Move the cursor to the upper middle portion of the screen and left click on **Dimension** as shown in Figure 15.

Figure 15   Left Click Here

22.   After selecting **Dimension** move the cursor over the bottom horizontal line. The line will turn red as shown in Figure 16. Select the line by left clicking anywhere on the line **or** on each of the end points. To use the end points of the line, move the cursor over one of the end points. A small red square will appear. Left click once and move the cursor to the other end point. Another red square will appear. Left click once. The dimension will now be attached to the cursor. Move the cursor up and down to verify it is attached.

Figure 16   Turned Red

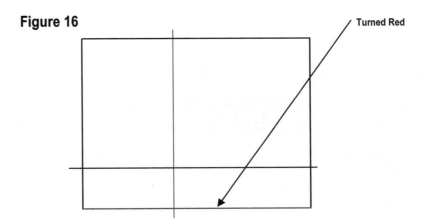

23.   Move the cursor down. The actual dimension of the line will appear as shown in Figure 17.

Figure 17   Left Click Here

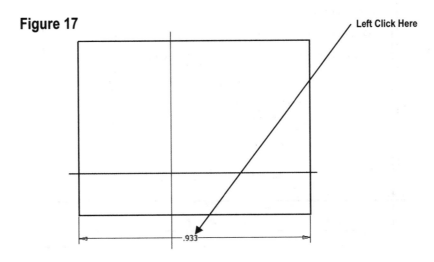

24. Move the cursor to where the dimension will be placed and left click once. While the dimension is still red, left click once. The Edit Dimension dialog box will appear as shown in Figure 18.

25. To edit the dimension, enter **2.00** in the Edit Dimension dialog box (while the current dimension is highlighted) and press **Enter** on the keyboard.

Figure 18

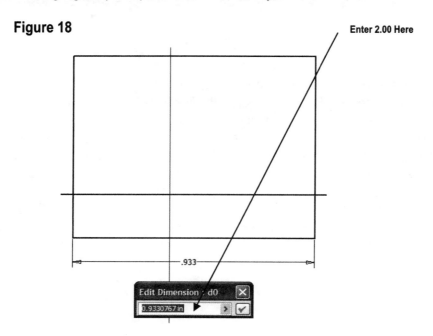

Enter 2.00 Here

.933

Edit Dimension : d0

0.9330767 in

26. The dimension of the line will become 2 inches as shown in Figure 19.

Figure 19

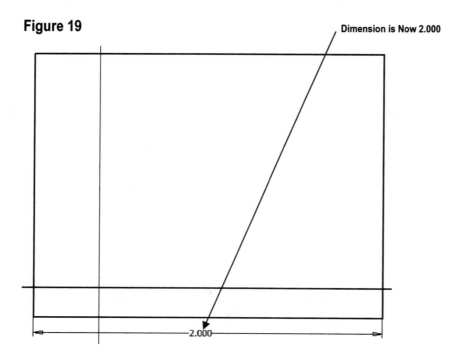

Dimension is Now 2.000

2.000

27. Move the cursor to the upper middle portion of the screen and left click on **Dimension** as shown in Figure 20.

Figure 20

Left Click Here

28. After selecting **Dimension** move the cursor over the right side vertical line. The line will turn red as shown in Figure 21. Left click once on the line.

Figure 21

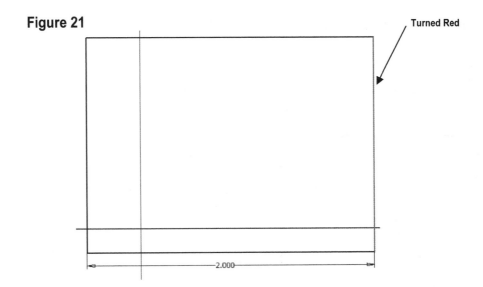

Turned Red

2.000

29. The dimension is attached to the cursor. Move the cursor up and down to verify it is attached. Move the cursor to the right of the line where the dimension will be placed and left click once. While the dimension is still red, left click the mouse once. The Edit Dimension dialog box will appear as shown in Figure 22.

Figure 22

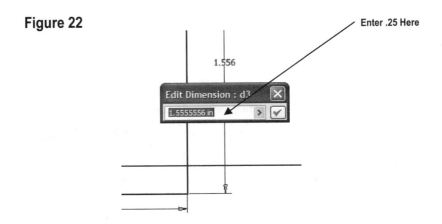

Enter .25 Here

1.556

Edit Dimension : d3

1.5555556 in

30. To edit the dimension, enter **.25** in the Edit Dimension dialog box (while the current dimension is highlighted) and press **Enter** on the keyboard.

31. The screen should look similar to Figure 23.

Figure 23

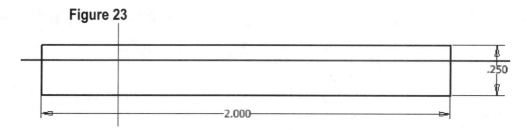

32. Complete the remainder of the sketch as shown in Figure 24.

Figure 24

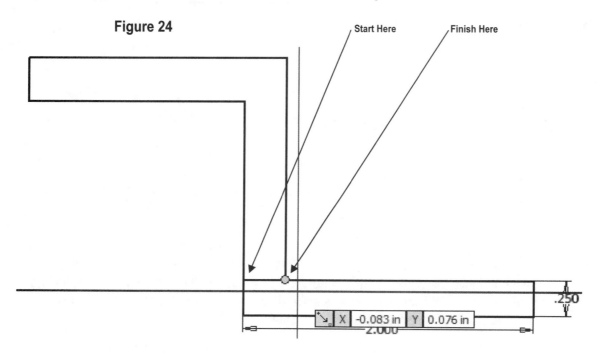

33. Move the cursor to the upper middle portion of the screen and left click on **Trim** as shown in Figure 25.

Figure 25

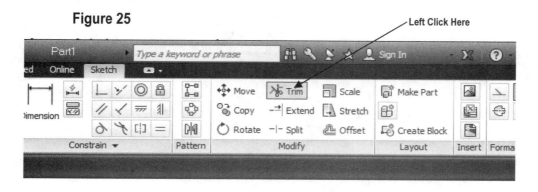

34. Move the cursor over the portion of the line that is shown in Figure 26. The line will become dashed. Inventor is guessing that this line will be trimmed.

Figure 26

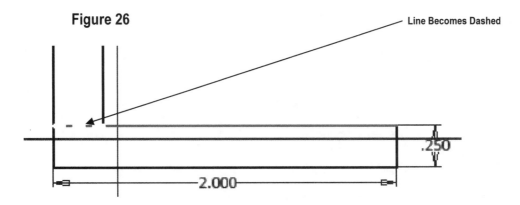

35. While the line is dashed, left click on the dashed portion. The line will be trimmed as shown in Figure 27.

Figure 27

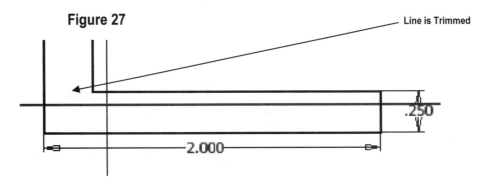

36. Move the cursor over the line in the lower left corner of the drawing as shown in Figure 28. The line will turn red. This particular line will have to be deleted so that the line above can be extended the full length.

Figure 28

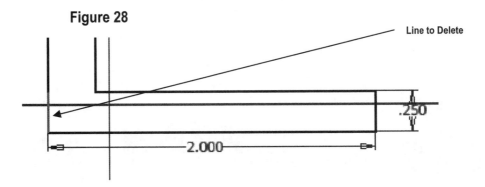

37. After the line turns red, right click the mouse. A pop up menu will appear. Left click on **Delete** as shown in Figure 29.

Figure 29

Left Click Here

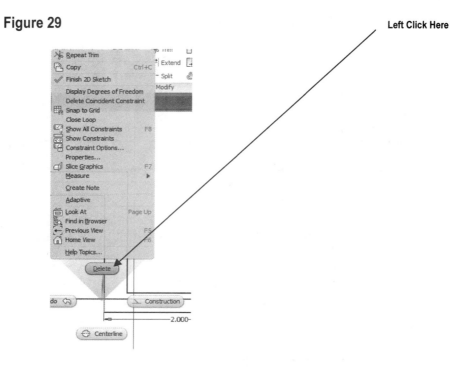

38. The line will be deleted as shown in Figure 30.

Figure 30

Line is Deleted

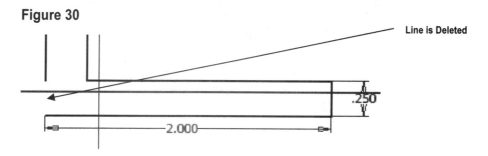

.250

2.000

39. Move the cursor to the upper middle portion of the screen and left click on **Extend** as shown in Figure 31.

Figure 31

Left Click Here

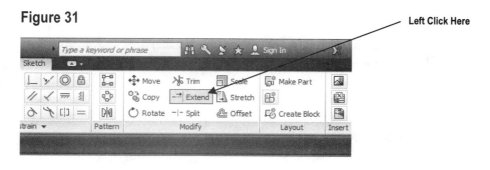

40. Move the cursor to the line above the recently deleted line. This is the line that will be extended. After the cursor is over the line it will turn red and extend a line downward, as shown in Figure 32.

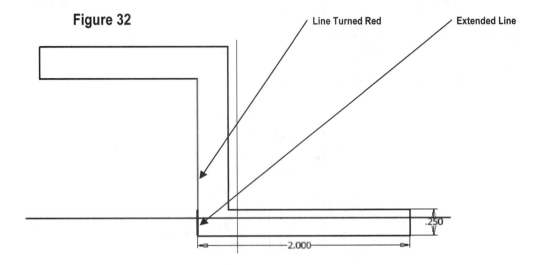

Figure 32

41. After the extended line appears, left click the mouse. The line will extend, creating one continuous line as shown in Figure 33. This is crucial in creating a sketch that can be extruded into a solid model

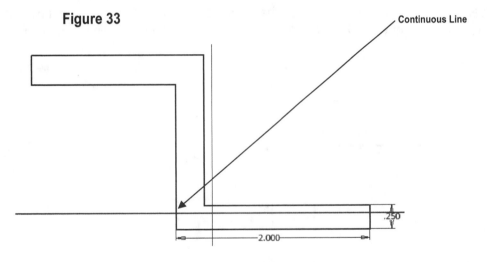

Figure 33

42. Move the cursor to the upper middle portion of the screen and left click on **Dimension** as shown in Figure 34.

Figure 34

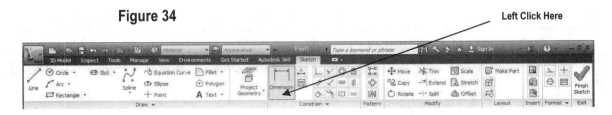

43. After selecting **Dimension** move the cursor over the left vertical line. The line will turn red as shown in Figure 35. Left click once on the line.

Figure 35

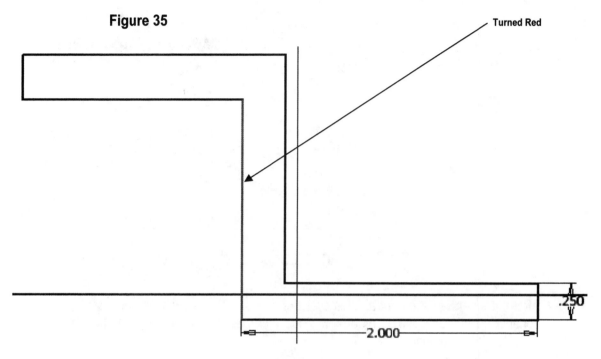

Turned Red

.250

2.000

44. Even though a dimension will be attached to the cursor simply ignore it and move the cursor to the right vertical line and left click after it turns red as shown in Figure 36.

Figure 36

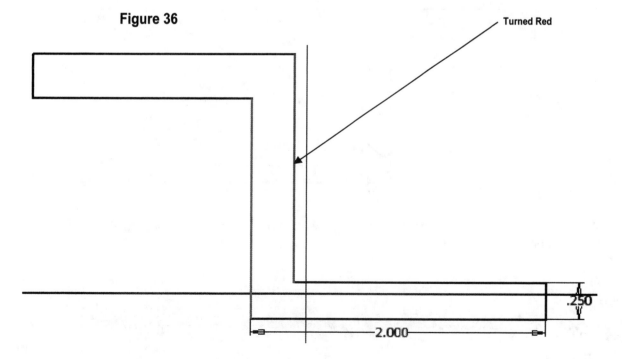

Turned Red

.250

2.000

45. Enter **.25** in the Edit Dimension dialog box (while the current dimension is highlighted) and press **Enter** on the keyboard.

46. Complete the remainder of the sketch as shown in Figure 37.

Figure 37

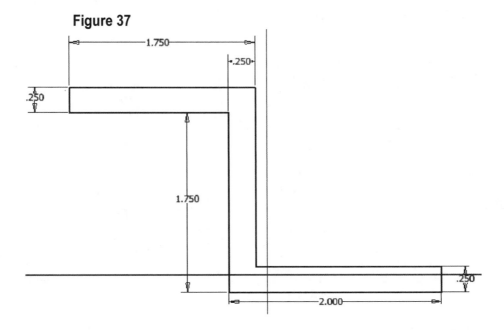

47. After the sketch is complete it is time to extrude the sketch into a solid. Right click anywhere around the sketch. A pop up menu will appear. Left click on **OK** as shown in Figure 38. Hitting the **ESC** key will also "get out" of the Dimension command.

Figure 38

Left Click Here

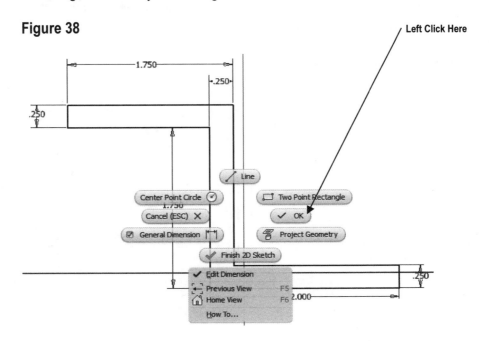

48.   After you have verified that no commands are active, right click anywhere on the sketch. A pop up menu will appear. Left click on **Finish 2D Sketch.** Left click **Home View** as shown in Figure 39.

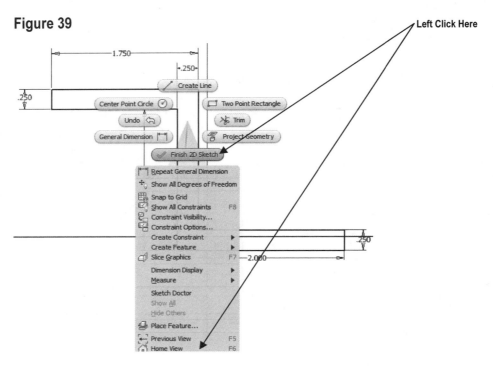

Figure 39

Left Click Here

49.   Inventor is now out of the Sketch Panel and into the Part Features Panel. Notice that the commands at the top of the screen are now different. To work in the Part Features Panel a sketch must be present and have no opens (non-connected lines). If there are any opens in the sketch an error message will appear. The view shown below is an Isometric/ Home view. Your screen should look similar to Figure 40.

Figure 40

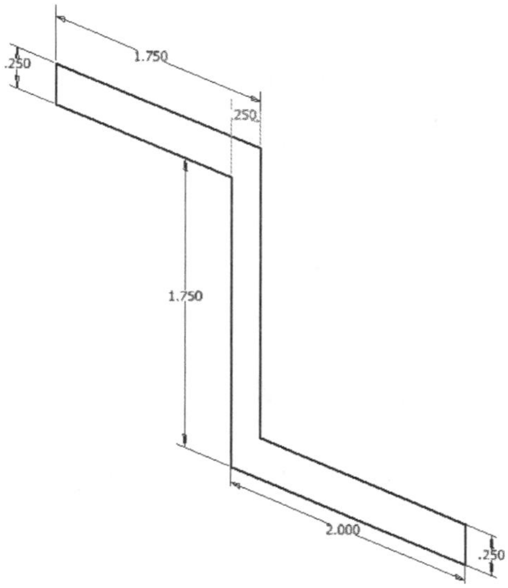

Extrude a sketch in the Part Features Panel using the Extrude command

50. Move the cursor to the upper left portion of the screen and left click on **Extrude.** The Extrude dialog box will appear in collapsed form. Left click on the drop down arrow in the center of the collapsed dialog box. This will cause the dialog box to expand. Inventor also provides a preview of the extrusion. If Inventor gave you an error message there are opens (non-connected lines) somewhere on the sketch. Check each intersection for opens by using the **Extend** and **Trim** commands. Your screen should look similar to Figure 41.

Figure 41

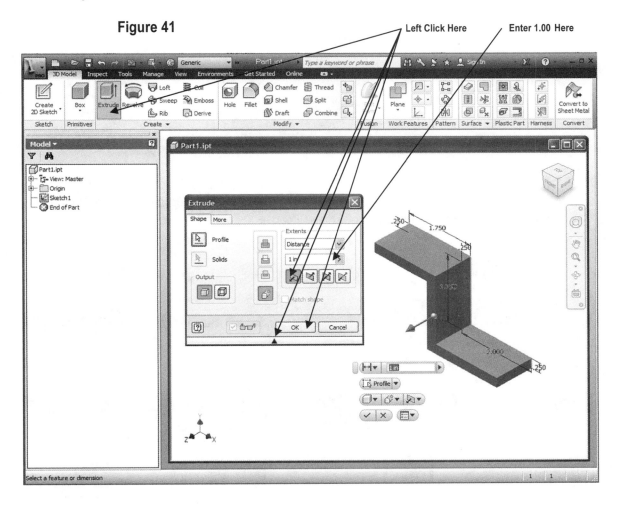

51. Enter **1.00** as shown and left click on the **OK**. Inventor will create a solid from the sketch as shown in Figure 41.

Create a fillet in the Part Features Panel using the Fillet command

52. Your screen should look similar to Figure 42.

Figure 42

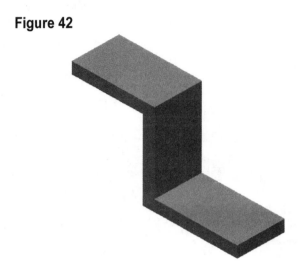

53. Move the cursor to the upper middle portion of the screen and left click on **Fillet**. The Fillet dialog box will appear in collapsed form. Left click on the drop down arrow in the center of the collapsed dialog box. This will cause the dialog box to expand as shown in Figure 43.

Figure 43

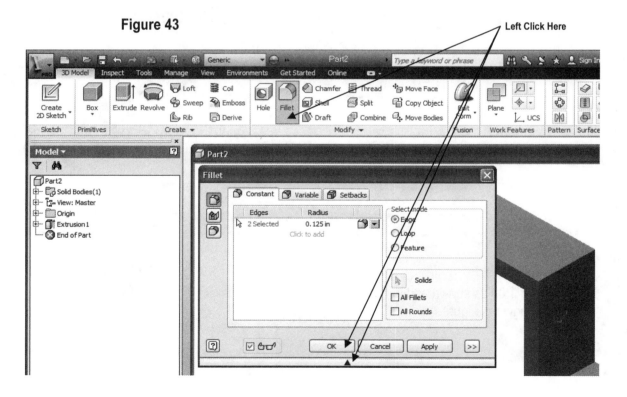

54. Move the cursor to the lower left edge of the part. After the edge turns red, left click once as shown in Figure 44.

Figure 44

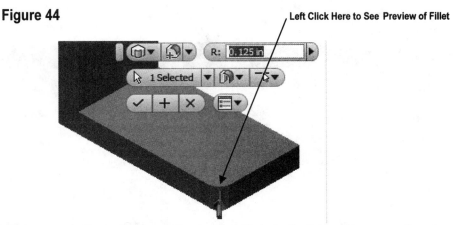

Left Click Here to See Preview of Fillet

55. Notice the blue mesh illustrating a preview of the fillet. Left click on the opposite edge as shown in Figure 45.

Figure 45

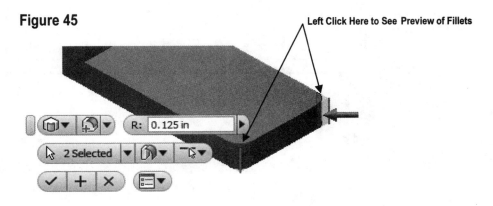

Left Click Here to See Preview of Fillets

56. Left click on the two upper remaining edges. Even though the far upper edge is not visible, move the cursor to the location of the edge and Inventor will find it as shown in Figure 46.

Figure 46

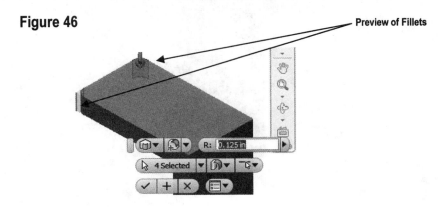

Preview of Fillets

57. Move the cursor to the dimension located in the dimension box. Enter **.5** and press **Enter** on the keyboard as shown in Figure 47.

Figure 47

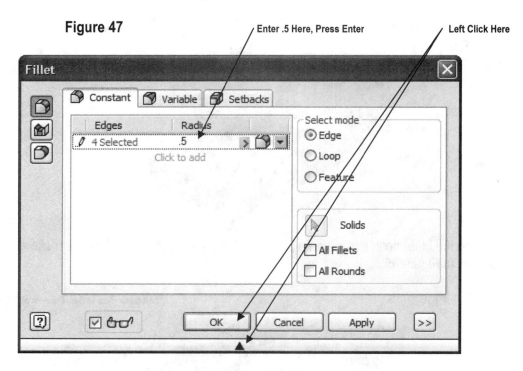

58. Your screen should look similar to Figure 48.

Figure 48

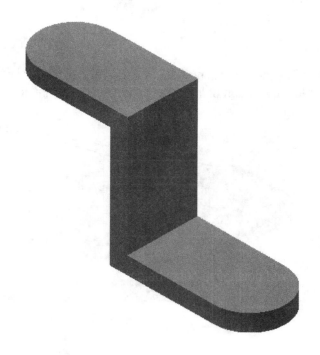

59. The next task will include cutting a hole in each of the ends. To accomplish this, a sketch will need to be constructed on this surface. Move the cursor to the surface that will have the new sketch as shown in Figure 49. Notice the edges of the surface become outlined in red.

Figure 49

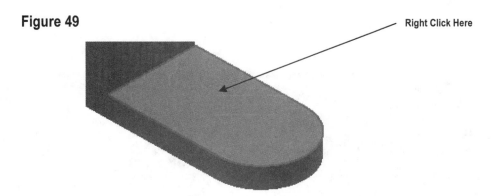

Right Click Here

60. After the edges of the surface turn red, right click on the surface. The surface will change color. A pop up menu will also appear. Left click on **New Sketch** as shown in Figure 50.

Figure 50

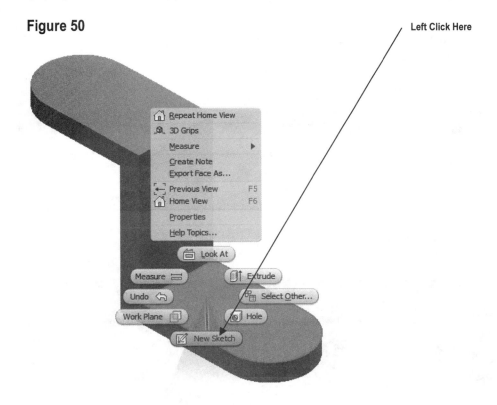

Left Click Here

61. Inventor will create a "sketch" on that particular surface. Notice the menu at the top of the screen has changed back to the options available in the Sketch Panel. Inventor has now returned to the sketch panel.

62. Your screen should look similar to Figure 51.

Figure 51

63. Move the cursor to the upper left portion of the screen and left click on **Circle** as shown in Figure 52.

Figure 52
Left Click Here

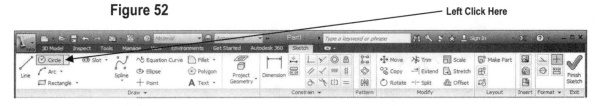

64. A green dot will appear at the center of the Fillet radius as shown in Figure 53.

Figure 53
Green Dot

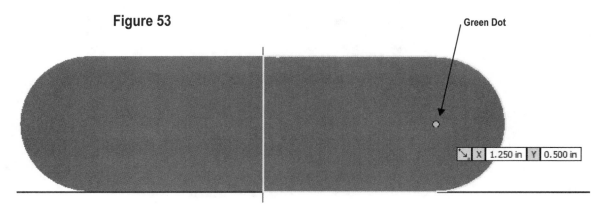

Create a hole in the Part Features Panel using the Extrude command

65. After the green dot appears, left click once. This will be the center of a circle, which will later become a thru hole. Move the cursor out to the side. The hole will become larger. Move the cursor out far enough to create a hole size similar to Figure 54.

Figure 54

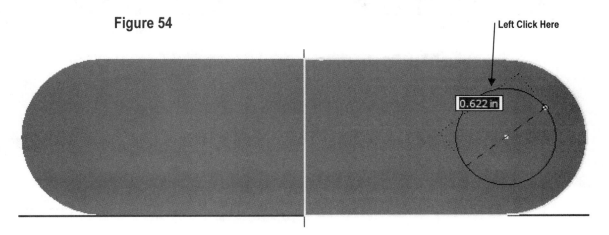

66. After the hole size looks similar to Figure 54, left click once.

67. Using the **Dimension** command, enter **.50** in the Edit Dimension dialog box and press **Enter** on the keyboard. The diameter of the hole will become .50 inches. Press the **Esc** key to ensure that no commands are active.

68. Right click anywhere on the drawing. A pop up menu will appear. Left click on **Finish 2D Sketch** or left click on **Finish Sketch/Exit** at the upper right portion of the screen. Left click on **Home View** as shown in Figure 55.

Figure 55

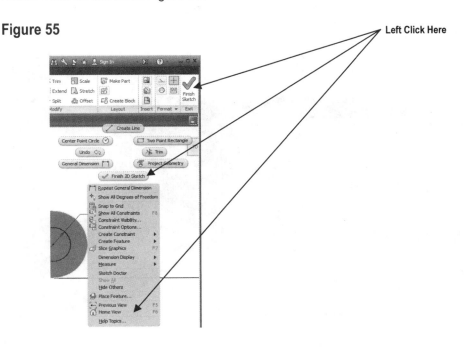

69. Inventor is now out of the Sketch Panel and into the Part Features Panel. Your screen should look similar to Figure 56.

Figure 56

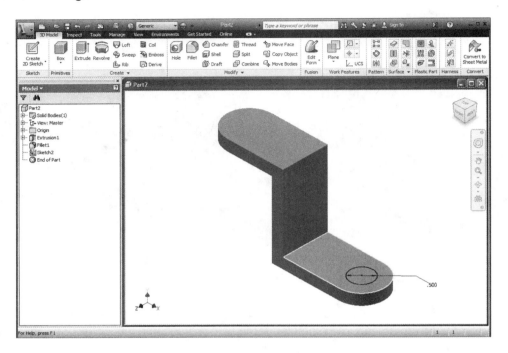

70. Move the cursor to the upper left portion of the screen and left click on **Extrude.** You may have to left click on the drop down arrow located in the lower center of the dialog box to expand it as shown in Figure 57. Now move the cursor over the circle in the drawing. After the circle turns red, left click once. Enter **.25** for the depth. Left click on the directional arrow to define the type of extrusion. Left click on **Cut**. Left click on **OK** as shown in Figure 57.

Figure 57

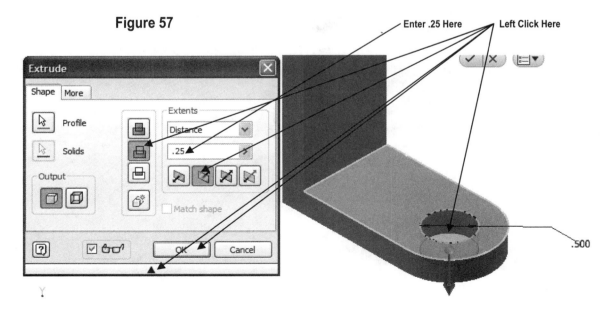

Create a counter bore in the Part Features Panel using the Hole command

71. Your screen should look similar to Figure 58.

Figure 58

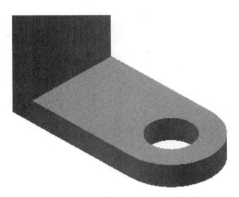

72. Another method of creating a hole is to use the Point, Center Point command.

73. To use the Point, Center Point command, Inventor will need to be in the Sketch Panel. Right click on the surface as shown in Figure 59.

Figure 59   Right Click Here

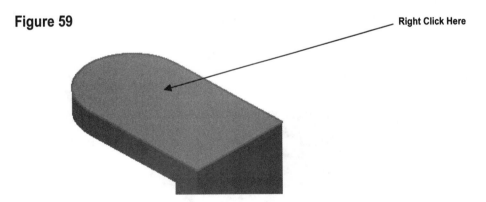

74. The surface will change color. A pop up menu will also appear. Left click on **New Sketch** as shown in Figure 60.

Figure 60   Left Click Here

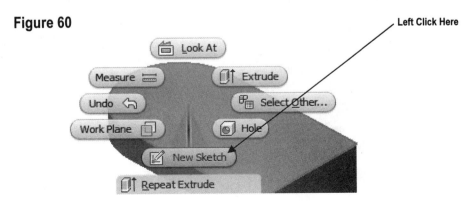

75. Inventor will return to the Sketch Panel as shown in Figure 61.

Figure 61

76. Move the cursor to the middle left portion of the screen and left click on **Point** as shown in Figure 62.

Figure 62   **Left Click Here**

77. A green dot will appear at the center of the Fillet radius. Left click on the green dot as shown in Figure 63.

Figure 63   **Left Click Here**

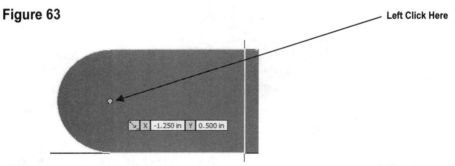

78. After left clicking on the center point, Inventor will place a small center marker on the center of the fillet radius as shown in Figure 64. Press the **Esc** key once.

Figure 64   **Center Marker**

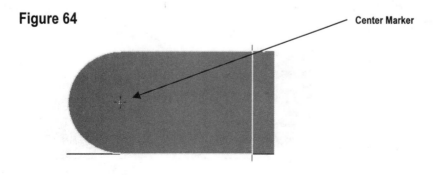

79. Right click anywhere on the drawing. A pop up menu will appear. Left click on **Finish 2D Sketch** as shown in Figure 65.

Figure 65

Left Click Here

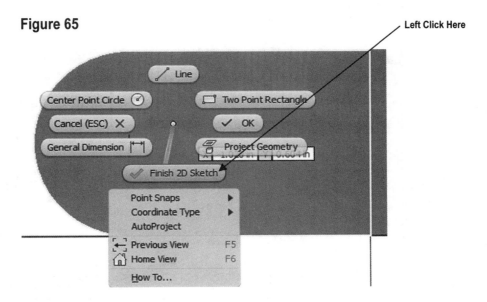

80. Inventor is now out of the Sketch Panel and into the Part Features Panel. Your screen should look similar to Figure 66.

Figure 66

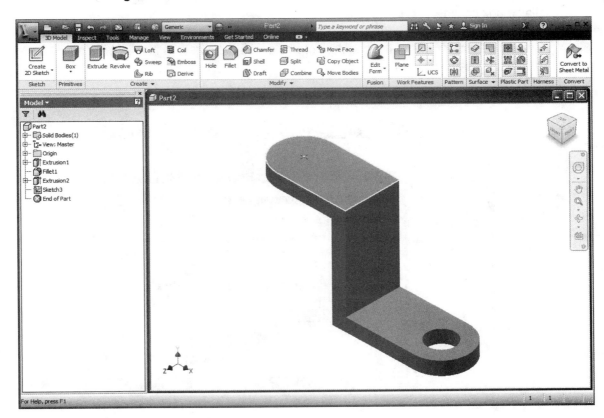

81.   Move the cursor to the upper middle portion of the screen and left click on **Hole**. The Hole dialog box will appear as shown in Figure 67.

Figure 67

Left Click Here

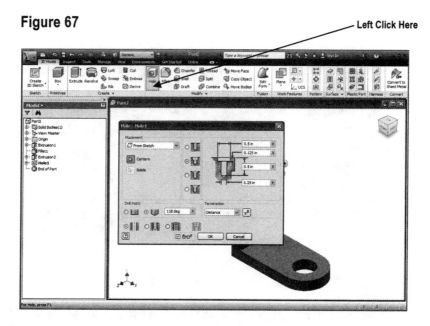

82.   Place a dot as shown in Figure 67. This is the "Counter Bore" icon. A preview and dimensions of the hole type are provided on the right side of the Holes dialog box. Select one of the other hole types and watch the preview of the hole in the right side of the Holes dialog box change.

83.   To edit the dimensions of the counter bore hole, use the cursor to highlight the desired dimension as shown in Figure 68.

Figure 68

Enter 0.50 Here

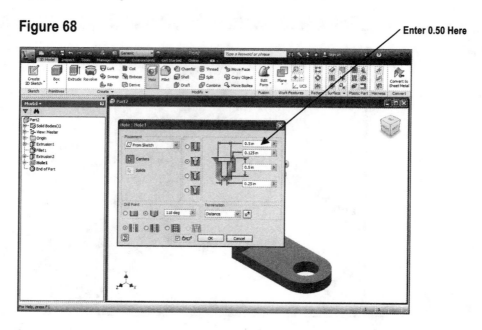

84. After highlighting the dimension, enter in **.50** (if .50 is not already entered in) for the counter bore diameter.

85. Enter in **.125** for the counter bore depth, **.50** for the overall depth, **.25** for the hole diameter. Left click on **Distance** under Termination and left click on **OK** as shown in Figure 69.

Figure 69

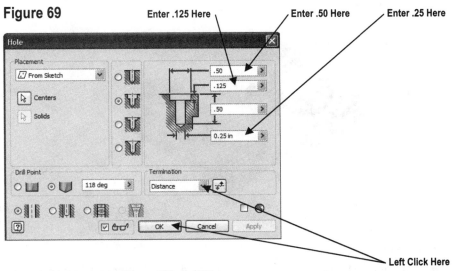

86. Your screen should look similar to Figure 70.

Figure 70

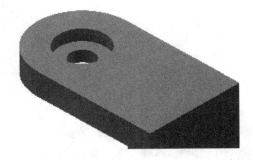

87. To ensure that the hole is correct, move the cursor to the upper left portion of the screen and left click on the **View** tab. Left click on the "Free Orbit/Orbit/Rotate" icon (previous version of Inventor display "Orbit or Rotate") as shown in Figure 71.

Figure 71

88. The Free Orbit/Rotate command will become active. Left click anywhere <u>inside</u> the white circle, hold the left mouse button down, and drag the cursor upward. The part will rotate upward as shown in Figure 72.

Figure 72

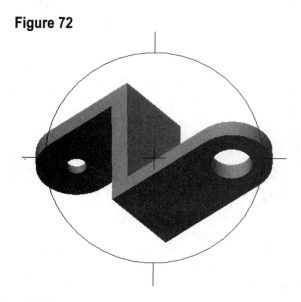

89. Holding the left mouse button down keeps the part attached to the cursor. To view the part in Isometric, right click anywhere on the screen and left click on **Home View** from the pop up menu as shown in Figure 73.

Figure 73

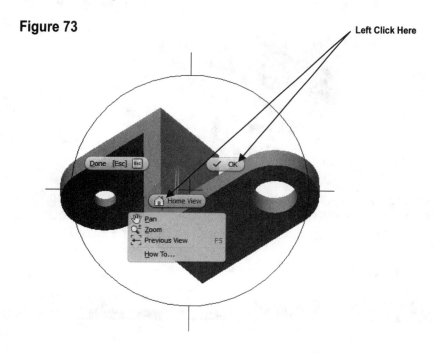

Left Click Here

90. As long as the white circle is present, the Free Orbit/Rotate command is still active. To get out of the Orbit/Rotate command either use the keyboard and press **Esc** once or twice, or right click anywhere on the screen. A pop up menu will appear. Left click on **OK** from the pop up menu shown in Figure 74.

Figure 74

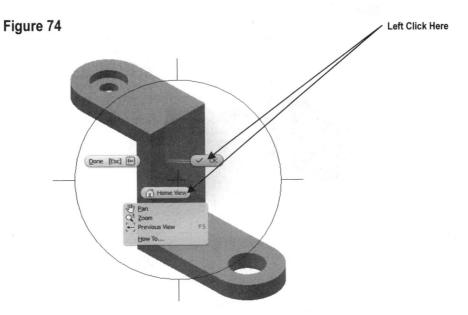

91. Other commands for viewing are at the upper right portion of the screen. Left click on the **View** tab to access each command. You can also use the icons located at the far right. Each icon has a drop down arrow to locate more viewing options. Left click on the drop down arrow under each command to become more familiar with what options are available as shown in Figure 75. Probably the most convenient tool for Zooming in and out is the mouse wheel. Simply roll the wheel forward to move the view away. To bring the view closer, simply roll the wheel towards you.

Figure 75

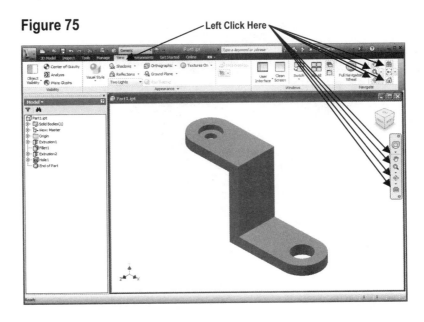

92. To save the model, move the cursor to the upper left portion of the screen and left click on the "**I**" at the far upper left. A drop down menu will appear as shown in Figure 76. Save the file where it can be retrieved later.

Figure 76

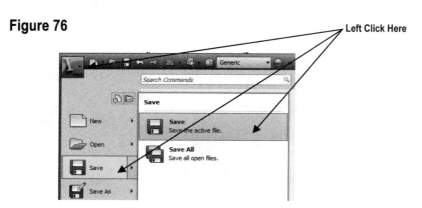

Chapter Problems

Use the Extrude and Extrude-Cut commands to complete the following.

Problem 1-1

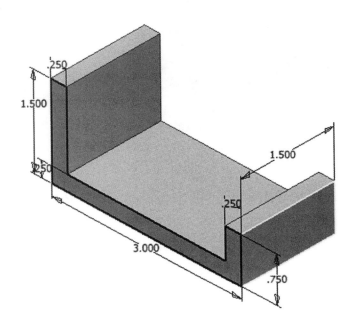

Problem 1-2

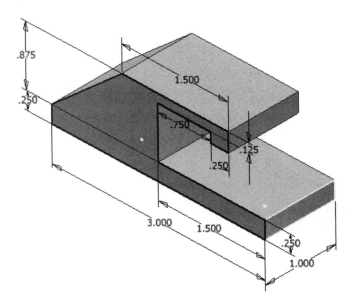

Problem 1-3

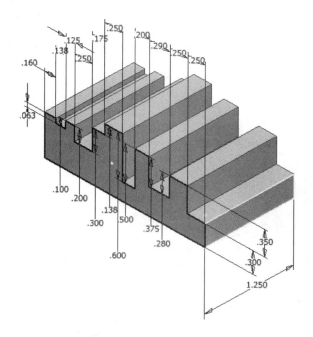

Problem 1-4

Hint: Use Extrusions to "Cut" features into the Part as shown.
L-Shaped Extrusion with Two (2) Negative Cuts

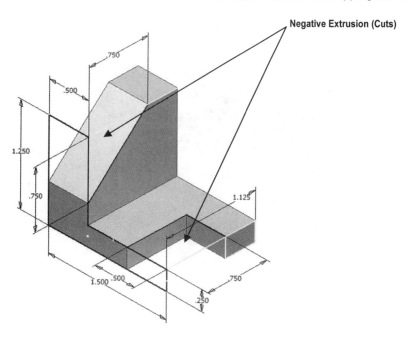

Problem 1-5

Hint: Use Extrusions to "Cut" features into the Part as shown.
H-Shaped Extrusion with One (1) Negative Cut

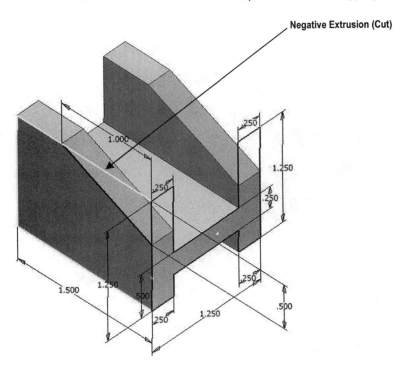

Problem 1-6

Hint: Use Extrusions to "Cut" features into the Part as shown.
Z-Shaped Extrusion with One (1) Negative Cut

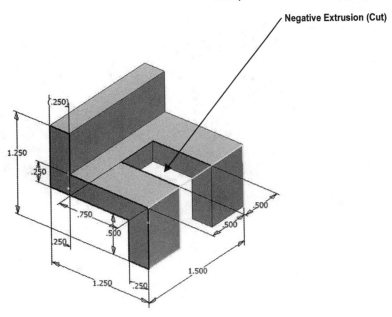

Problem 1-7

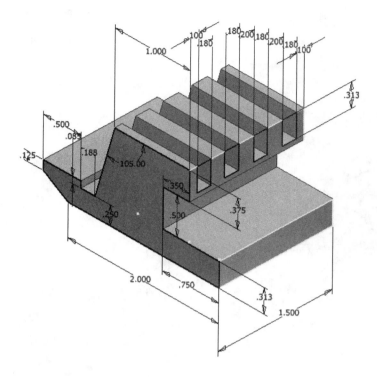

Problem 1-8

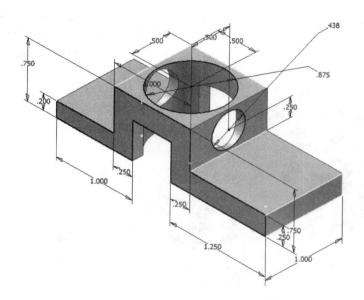

Problem 1-9

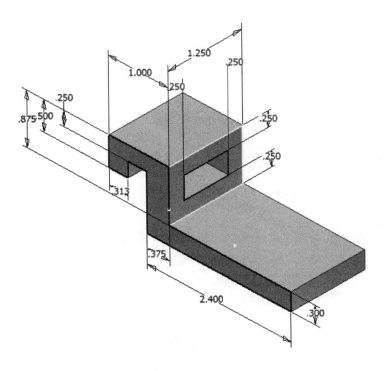

Problem 1-10

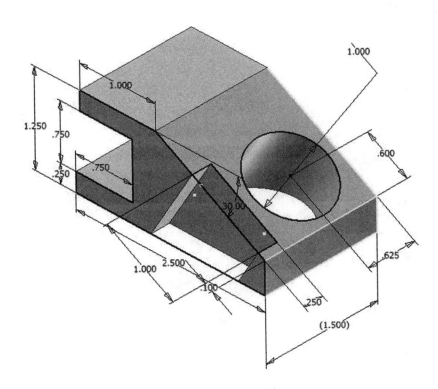

Problem 1-11

Hint: Create a Point while in the Sketch Panel to use the Hole Wizard
Counterbore Diameter .50 x .125 Deep
Thru Hole Diameter .25

Point

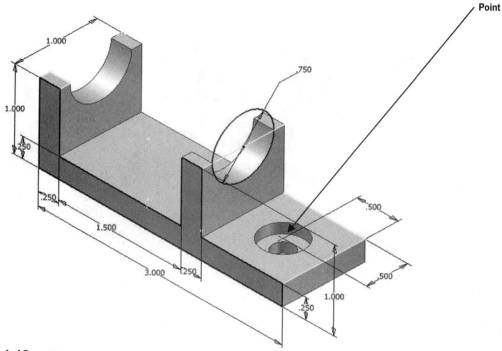

Problem 1-12

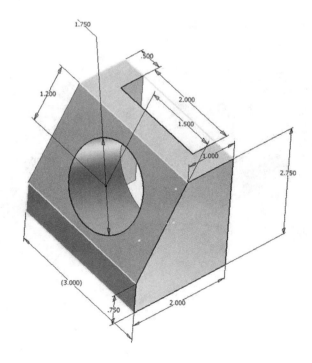

Problem 1-13

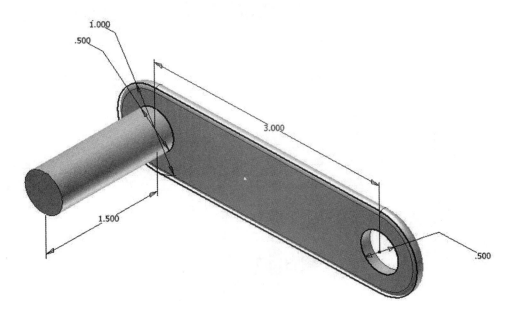

Problem 1-14

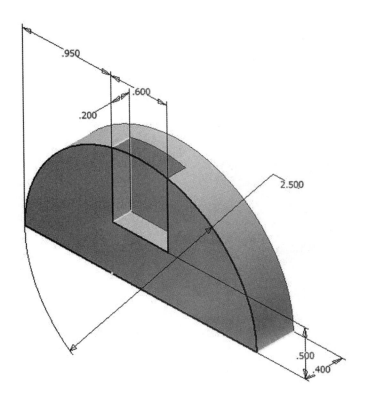

Problem 1-15

Hint: (2) Counterbored Holes
.75 Diameter x .375 Deep, Thru Hole Diameter = .375
Fillet Radius .375

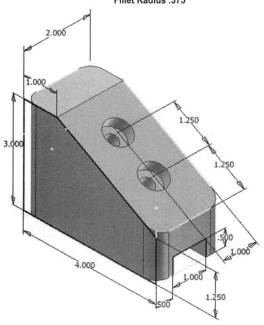

Problem 1-16

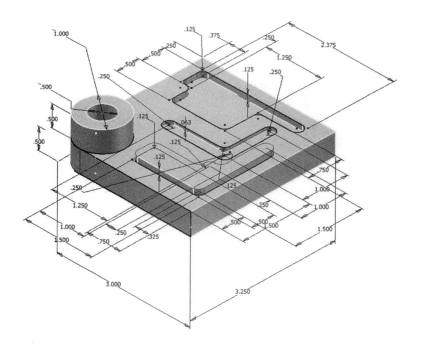

Learning
More Basics

Objectives:

1. Create a simple sketch using the Sketch Panel
2. Dimension a sketch using the Dimension command
3. Revolve a sketch in the Model/Part Features Panel using the Revolve command
4. Create a groove using the Revolve Cut command
5. Create a hole in the Model/Part Features Panel using the Extrude cut command
6. Create a series of holes in the Model/Part Features Panel using the Circular Pattern and command

Chapter 2 includes instruction on how to design the part shown.

1. Start Autodesk Inventor 2014 by referring to "Chapter 1 Getting Started".

2. After Autodesk Inventor 2014 is running, begin a new sketch.

3. Complete the sketch (including dimensions) as shown. Include the line above the sketch as shown in Figure 1. This line will be used to revolve the sketch around.

Figure 1

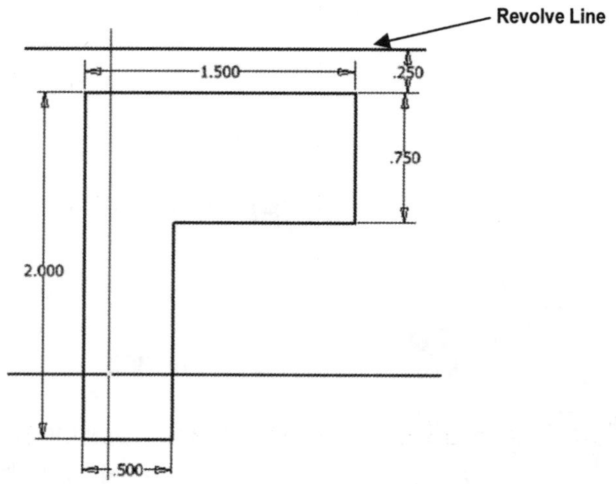

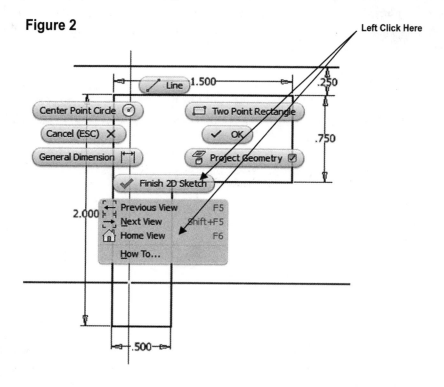

4. Right click around the sketch. A pop up menu will appear. Left click on **Home View**, then **Finish 2D Sketch** as shown in Figure 2.

Figure 2

5. The view will become isometric as shown in Figure 3.

Figure 3

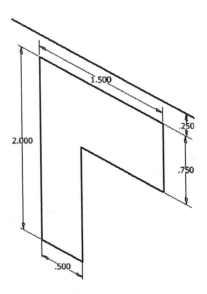

Revolve a sketch in the Part Features Panel using the Revolve command

6. Move the cursor to the upper left portion of the screen and left click on **Revolve**. Left click on the drop down arrow in the center of the collapsed dialog box. This will cause the dialog box to expand. If Inventor gave you an error message, there are opens (non-connected lines) somewhere on the sketch OR the view is not Isometric. Check each intersection for opens by using the **Extend** and **Trim** commands and make sure the view is Isometric. Your screen should look similar to Figure 4.

Figure 4

Left Click Here

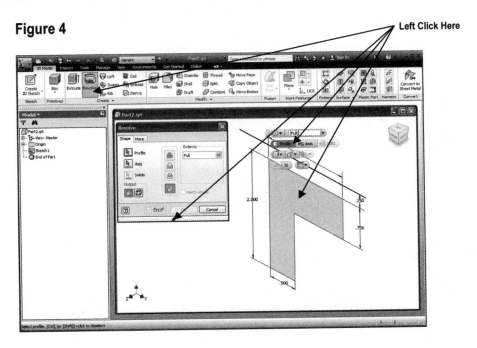

7.   Notice that the Profile icon has already been selected. Because there is only one profile present, Inventor assumes that particular profile will be selected. If the drawing contained more than one profile, you would have to first select the profile icon in the revolve dialog box then use the cursor to select the desired profile.

8.   Left click on the **Axis** icon. Move the cursor over the axis causing it to turn red and left click once as shown in Figure 5.

Figure 5

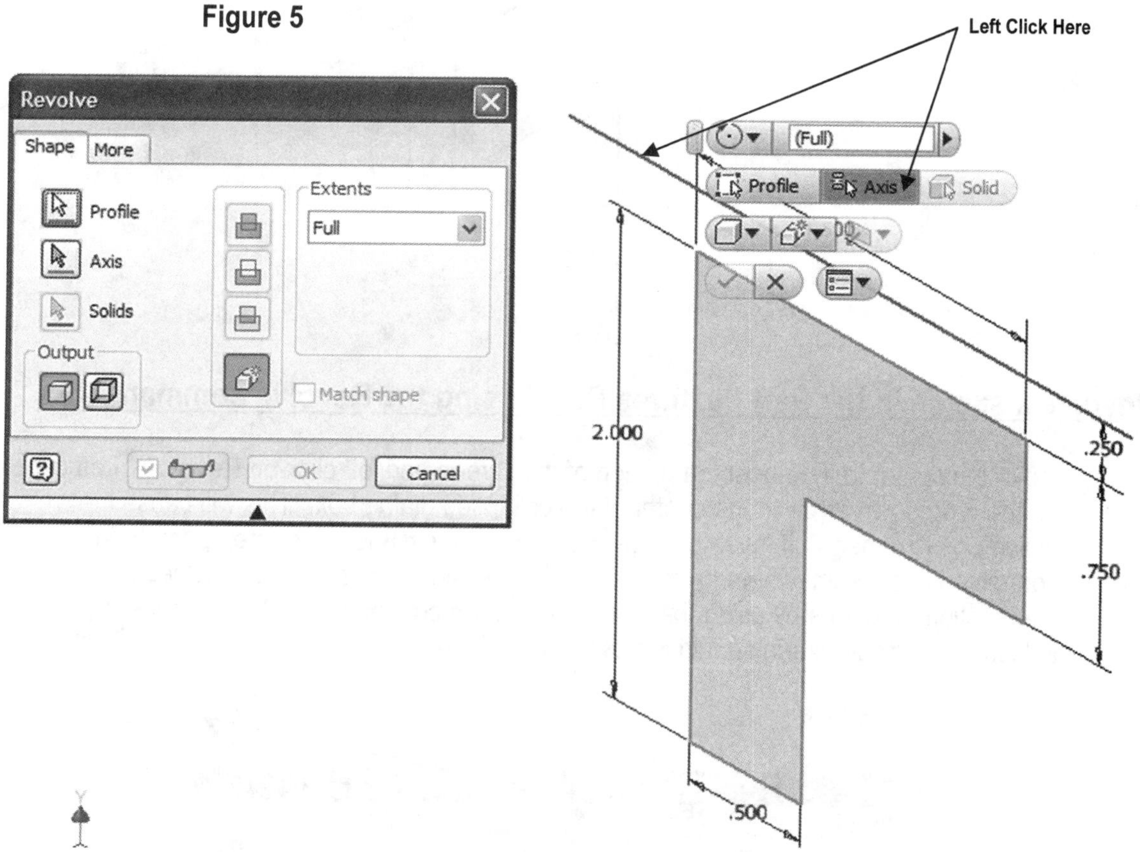

9.   A preview of the revolve will appear as shown in Figure 6.

Figure 6

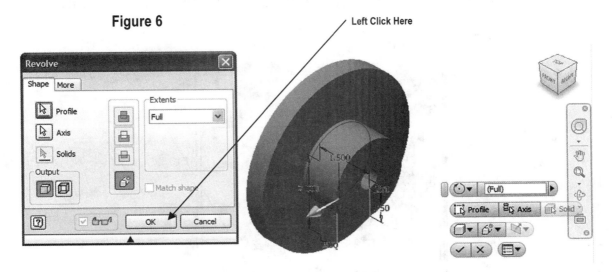

10.   Left click on **OK**.

11.   Your screen should look similar to Figure 7. You may have to zoom out to view the entire part.

Figure 7

12. Move the cursor over "XY Plane" located in the upper left portion of the screen and right click once. A pop up menu will appear. Left click on **New Sketch** as shown in Figure 8.

Figure 8

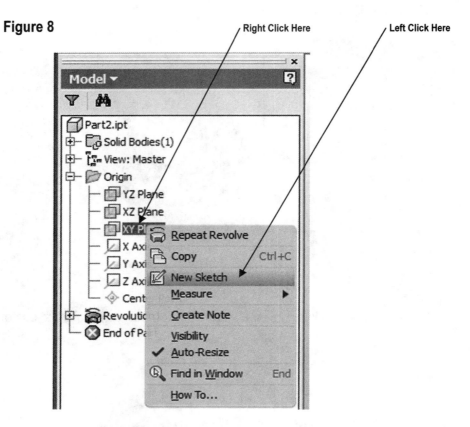

Right Click Here   Left Click Here

13. Inventor will create a work plane in the center of the part. You may have to select **Home View** to see the plane as shown in Figure 9.

Figure 9

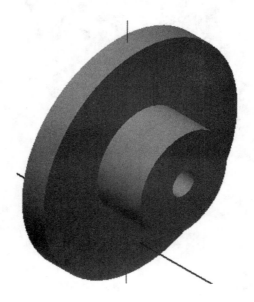

14. Move the cursor to the upper middle portion of the screen and left click on the **View** tab. Left click on the arrow under "Visual Style". A drop down menu will appear. Left click on **Wireframe** as shown in Figure 10.

Figure 10

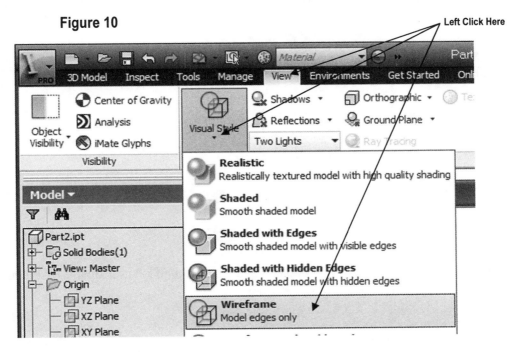

15. Inventor will change the display of the model to wire frame as shown in Figure 11.

Figure 11

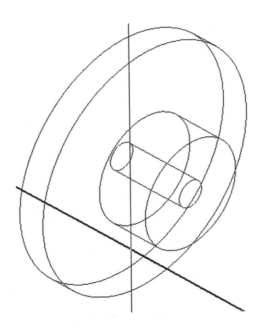

16. Move the cursor to the middle right portion of the screen and left click on the "Face View/ Look At" icon as shown in Figure 12.

Figure 12

Left Click Here

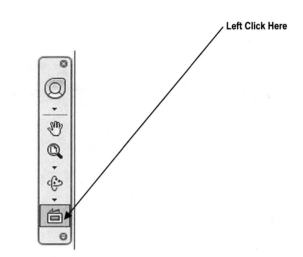

17. Move the cursor to the upper left portion of the screen and left click on the XY Plane as shown in Figure 13.

Figure 13

Left Click Here

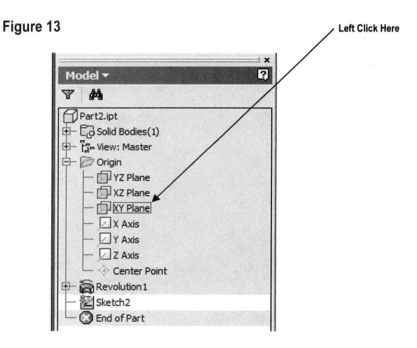

18.   Inventor will rotate the model around providing a perpendicular view of the XY plane as shown in Figure 14.

Figure 14

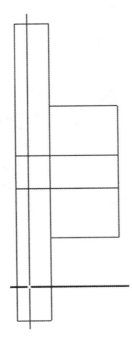

19.   Use the Free Orbit/Rotate command to rotate the part slightly from perpendicular as shown in Figure 15.

Figure 15

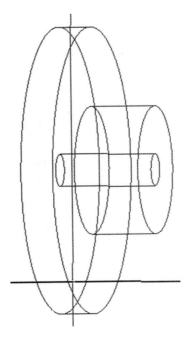

Use the Revolve Cut command to create a groove

20. Move the cursor to the upper middle portion of the screen and left click on **Project Geometry.** You may need to left click on the **Sketch** tab if the Project Geometry icon is not visible as shown in Figure 16.

Figure 16

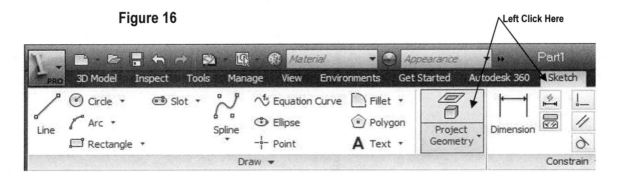

21. Move the cursor of the left side line. When it becomes highlighted (turns red) left click on it once as shown in Figure 17.

Figure 17

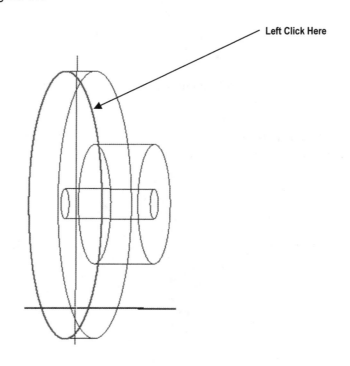

22.   Inventor will Project this line on to the New Sketch as shown in Figure 18.

Figure 18

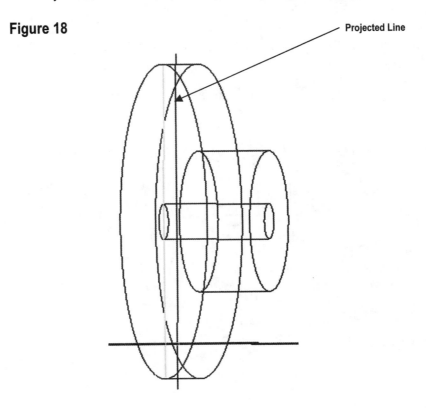

23.   Move the cursor to the upper middle portion of the screen and left click on **Project Geometry** as shown in Figure 19.

Figure 19

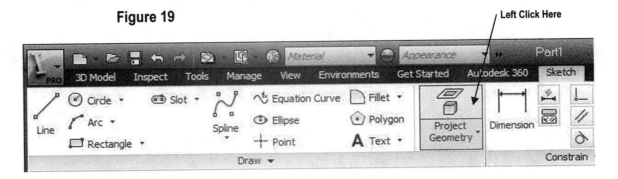

24. Move the cursor of the right side line. When it becomes highlighted (turns red) left click on it once as shown in Figure 20.

Figure 20

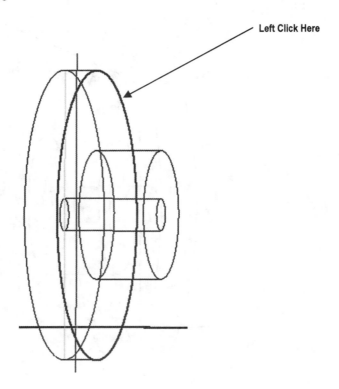

Left Click Here

25. Inventor will project this line on to the New Sketch as shown in Figure 21.

Figure 21

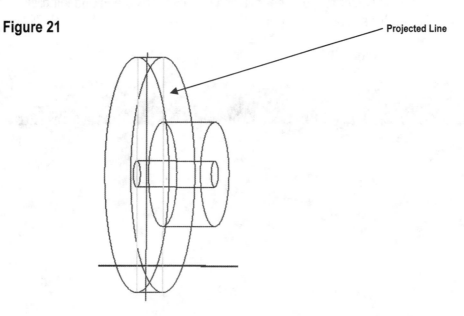

Projected Line

26.   Move the cursor to the upper portion of the screen and left click on the drop down arrow under Project Geometry. Left click on **Project Cut Edges** as shown in Figure 22.

Figure 22

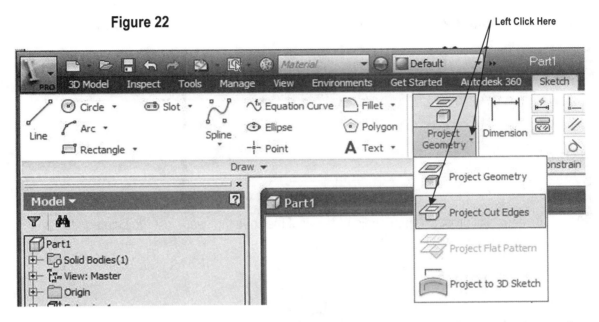

27.   Inventor will project all cut edges on to the new sketch as shown in Figure 23. Due to the image being in Grayscale for production purposes, all lines are not visible in Figure 23.

Figure 23

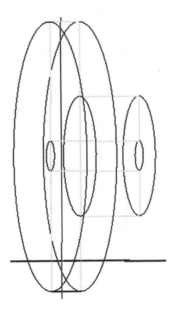

28. Move the cursor to the upper right portion of the screen and left click on **Front** as shown in Figure 24.

Figure 24

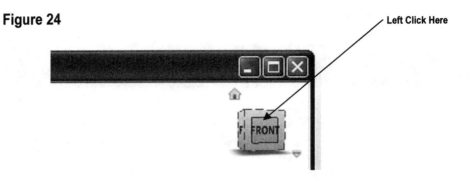

Left Click Here

29. Inventor will provide a perpendicular view of the lines that were just projected on to the sketch as shown in Figure 25.

Figure 25

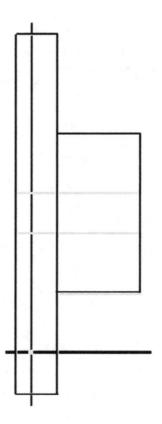

30. Complete the sketch as shown (draw a triangle) at the bottom of the part. Once the triangle is complete, delete all of the projected lines (including ones not shown) as shown in Figure 26.

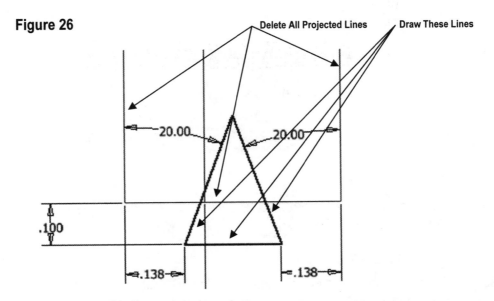

Figure 26

31. Draw an axis line/revolve line in the center of the part. Finish deleting all projected lines. Once deleted, projected lines may still appear in yellow. This is not a problem as shown in Figure 27.

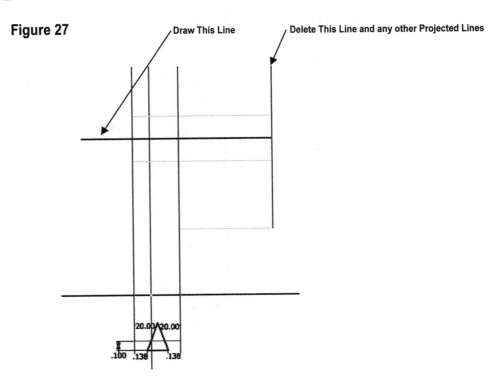

Figure 27

32. When deleting projected lines, if the option to delete a projected line does not appear in the pop up menu, then the projected line is already deleted even though it still appears in yellow as shown in Figure 28.

Figure 28

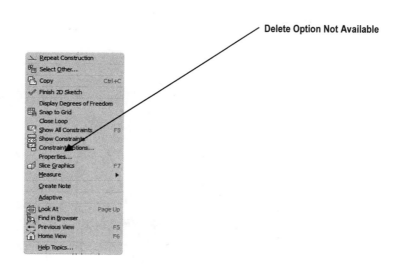

Delete Option Not Available

33. Exit out of the Sketch Panel using the Finish 2D Sketch command. Use the Orbit/Rotate command to rotate the part off to the side. Your screen should look similar to Figure 29.

Figure 29

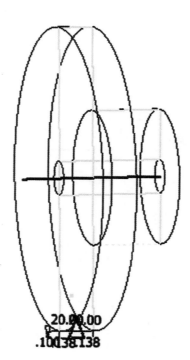

34. Move the cursor to the upper left portion of the screen and left click on the **3D Model** tab. Left click on **Revolve** as shown in Figure 30.

Figure 30

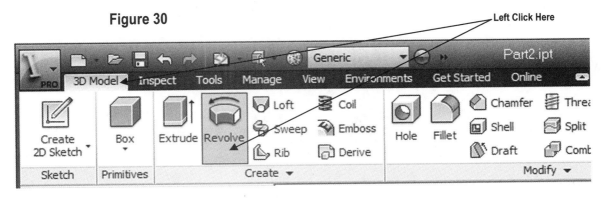

35. The Revolve dialog box will appear. Left click on the Profile icon and then on the small triangle created while in the Sketch Panel as shown in Figure 31.

Figure 31

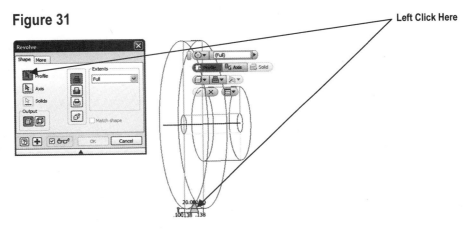

36. Left click on the Axis icon. Left click on the axis you created. Left click on the "Cut" icon. Left click on **OK** as shown in Figure 32.

Figure 32

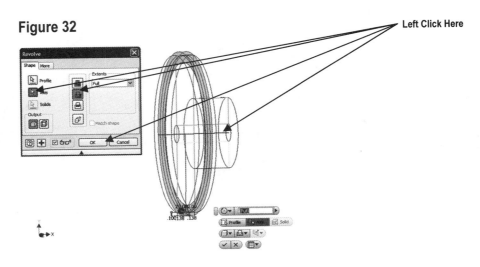

37.   Your screen should look similar to Figure 33.

Figure 33

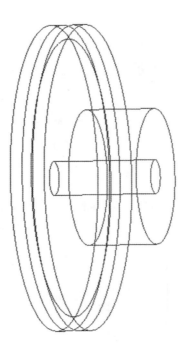

38.   Move the cursor to the upper middle portion of the screen and left click on the **View** tab. Left click on the arrow under "Visual Style". A drop down menu will appear. Left click on **Wireframe** as shown in Figure 34.

Figure 34

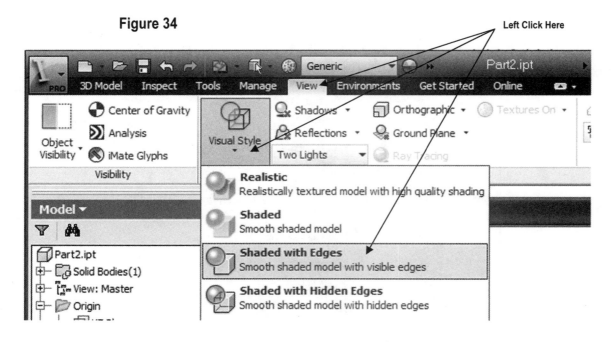

Create a hole in the Part Features Panel using the Extrude command

39.   Your screen should look similar to Figure 35.

Figure 35

40.   Rotate the part around using the **Home View** command. Move the cursor to the surface of the part causing the edge to turn red. After the edge becomes red, right click on the surface as shown in Figure 36.

Figure 36

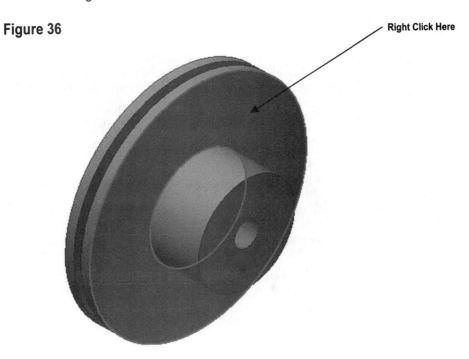

Right Click Here

41. The surface will turn blue and a pop up menu will appear. Left click on **New Sketch** as shown in Figure 37.

Figure 37

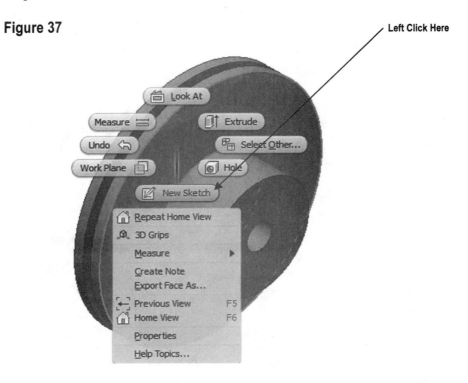

42. Inventor will begin a new sketch on the selected surface. Inventor will turn the view to the Front. You will need to use the **Home View** command to provide an isometric view. Your screen should look similar to Figure 38.

Figure 38

43. To gain a better look at the selected surface, move the cursor to the top center portion of the screen and left click on the **View** tab. Left click on the **Face View/Look At** icon. You can also left click on the "Right" portion of the View Cube as shown in Figure 39. The View Cube can also be used to rotate the part around (by holding the left mouse button down) similar to using the Orbit/Rotate icon.

Figure 39

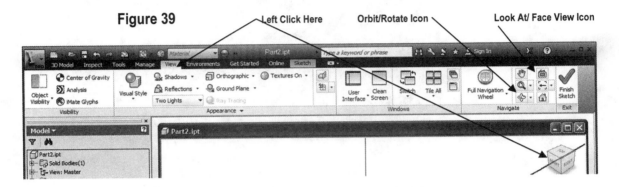

44. Left click on the surface the new sketch will be constructed on as shown in Figure 40.

Figure 40

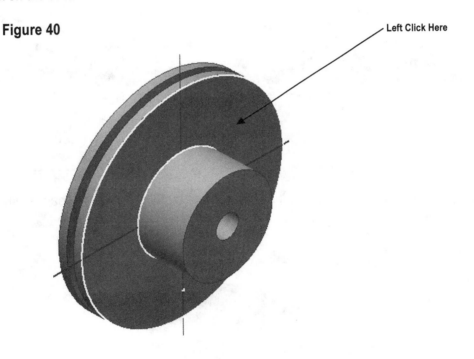

45. Inventor will rotate the part to provide a perpendicular view of the selected surface as shown in Figure 41.

Figure 41

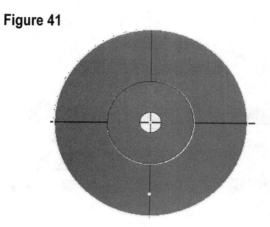

46. Move the cursor to the upper middle portion of the screen and left click on the **Sketch** tab. Left click on **Line** as shown in Figure 42.

Figure 42

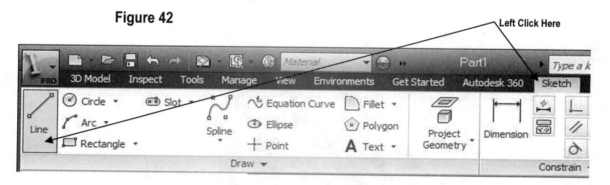

47. Left click on the center of the hole. Ensure that a green dot appears as shown in Figure 43.

Figure 43

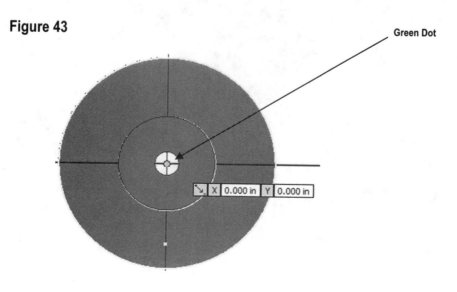

48. Move the cursor straight up and left click as shown in Figure 44.

Figure 44

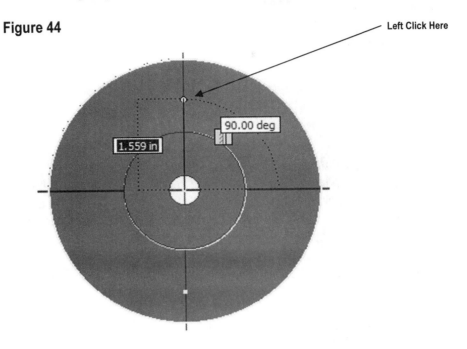

Left Click Here

49. Right click. A pop up menu will appear. Left click on **OK** as shown in Figure 45.

Figure 45

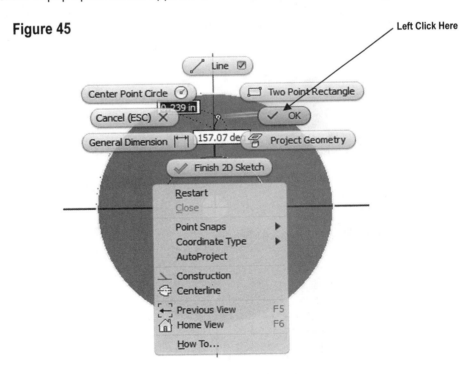

Left Click Here

50. Move the cursor to the upper middle portion of the screen and left click on **Dimension** as shown in Figure 46.

Figure 46

Left Click Here

51. After selecting **Dimension** move the cursor to the line that was just drawn. The line will turn red as shown in Figure 47. Select the line by left clicking anywhere on the line **or** on each of the end points. To use the end points of the line, move the cursor over one of the end points. A small red square will appear. Left click once and move the cursor to the other end point. After the red square appears, left click once. The dimension will be attached to the cursor.

Figure 47

Turned Red

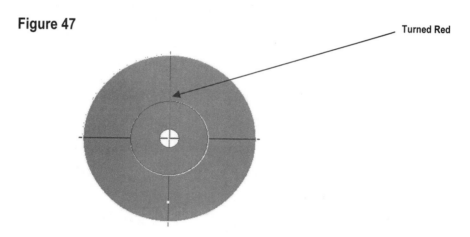

52. Move the cursor to the side. The actual dimension of the line will appear as shown in Figure 48

Figure 48

Left Click Here

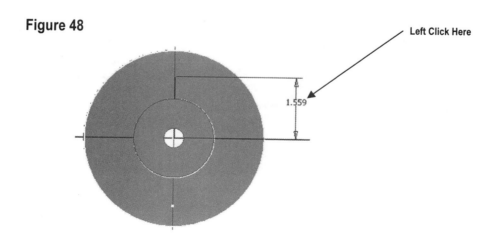

1.559

53. Move the cursor to where the dimension will be placed and left click once. While the dimension is still in red, left click once. The Edit Dimension dialog box will appear as shown in Figure 49.

Figure 49

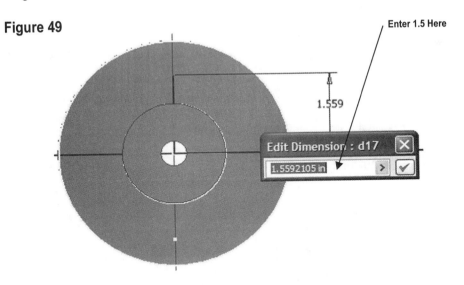

54. To edit the dimension, Enter **1.5** in the Edit Dimension dialog box (while the current dimension is highlighted) and press **Enter** on the keyboard.

55. The dimension of the line will become 1.5 inches as shown in Figure 50. Use the Zoom icons to zoom out if necessary.

Figure 50

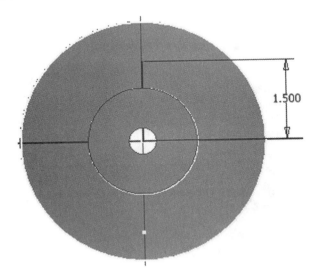

56. Move the cursor to the upper left portion of the screen and left click on **Circle** as shown in Figure 51.

Figure 51

57. Left click on the end point of the line as shown in Figure 52.

Figure 52

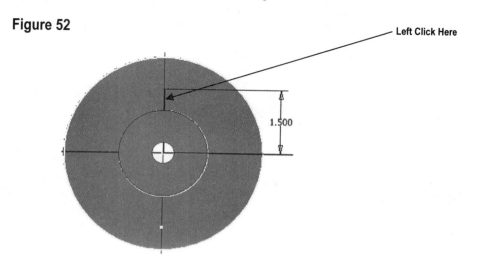

58. Move the cursor out to the right to create a circle as shown in Figure 53.

Figure 53

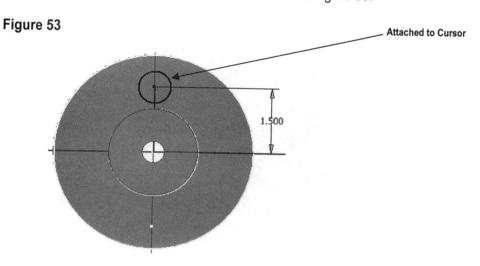

59.   Left click as shown in Figure 54. Press the **Esc** key once.

Figure 54

Left Click Here

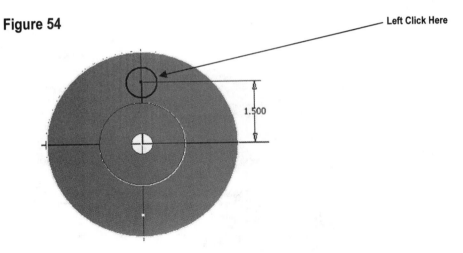

60.   Move the cursor to the upper middle portion of the screen and left click on **Dimension** as shown in Figure 55.

Figure 55

Left Click Here

61.   After selecting **Dimension** move the cursor to the edge of the circle that was just drawn. The circle will turn red.   Select the circle by left clicking anywhere on the circle (not the center) as shown in Figure 56. The dimension will be attached to the cursor.

Figure 56

Left Click Here

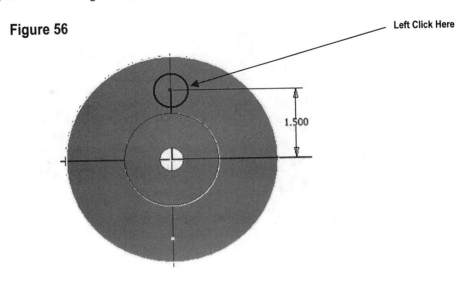

62.   Move the cursor to the side. The actual dimension of the line will appear as shown in Figure 57.

Figure 57

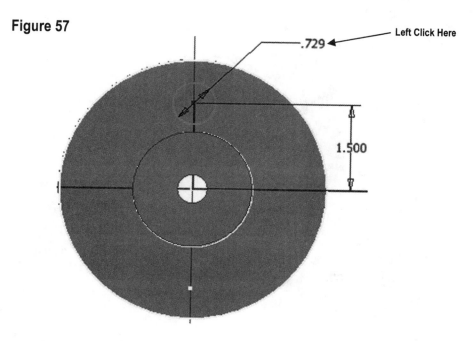

63.   Move the cursor to where the dimension will be placed and left click once. While the dimension is still in red, left click once. The Edit Dimension dialog box will appear as shown in Figure 58.

Figure 58

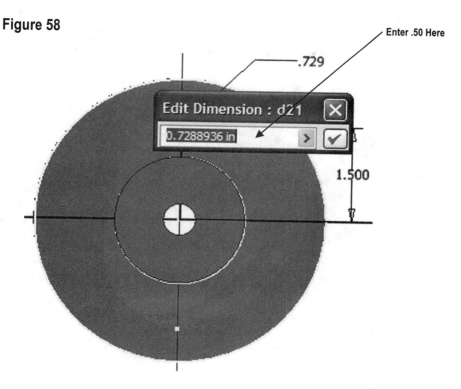

64. To edit the dimension, Enter **.50** in the Edit Dimension dialog box (while the current dimension is highlighted) and press **Enter** on the keyboard. Press the **Esc** key once or twice.

65. The dimension of the line will become .50 inches as shown in Figure 59. Use the Zoom icons to zoom out if necessary.

Figure 59

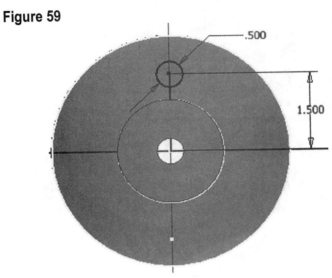

66. Move the cursor to the line that was used to locate the center of the circle. The line will turn red as shown in Figure 60.

Figure 60

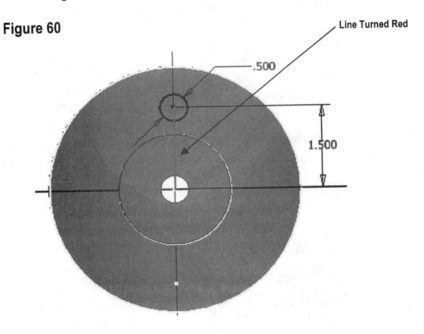

67. Right click on the line once it turns red. A pop up menu will appear. Left click on **Delete** as shown in Figure 61.

Figure 61

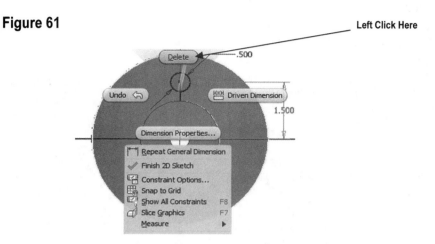

68. Press **Esc** once or twice, or right click around the drawing. A pop up menu will appear. Left click on **OK** or **Cancel (Esc)** as shown in Figure 62.

Figure 62

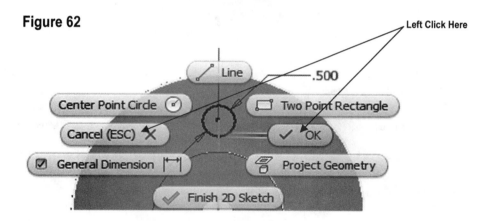

69. After you have verified that no commands are active, right click anywhere on the sketch. A pop up menu will appear. Left click on **Finish 2D Sketch** as shown in Figure 63.

Figure 63

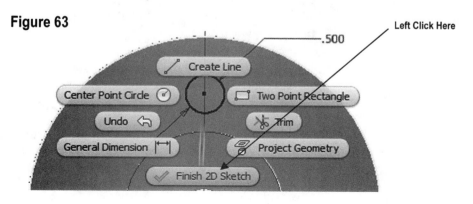

70.   Inventor is now out of the Sketch Panel and into the Part Features Panel. Notice that the commands at the top of the screen are now different. Your screen should look similar to Figure 64.

Figure 64

71.   Right click around the part. A pop up menu will appear. Left click on **Home View** as shown in Figure 65.

Figure 65

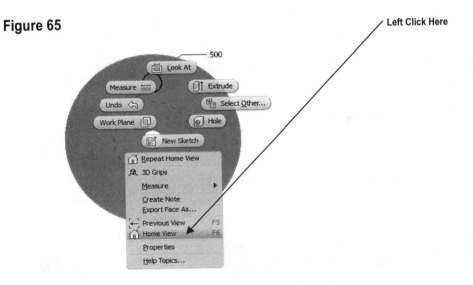

72.   The view will become Isometric as shown in Figure 66.

Figure 66

73. Move the cursor to the upper left portion of the screen and left click on **Extrude**. The Extrude dialog box will appear as shown in Figure 67.

Figure 67

Left Click Here

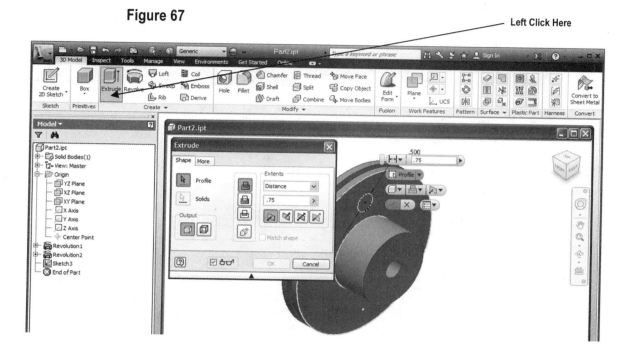

74. Move the cursor to the inside of the circle causing it to turn red as shown in Figure 68.

Figure 68

Move Cursor Here

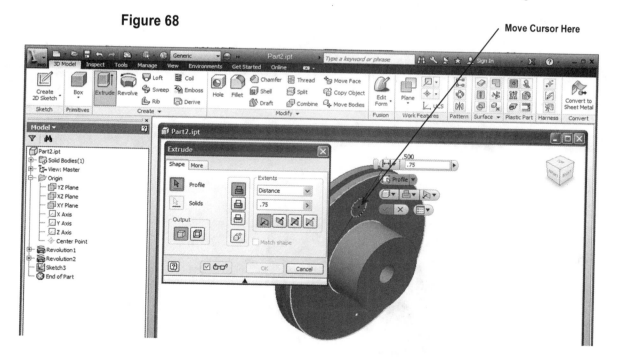

75. After the hole turns red, left click once. Select the "Cut" icon in the Extrude dialog box. Select the "Direction" icon to ensure the extrusion occurs in the right direction and left click on **OK** as shown in Figure 69.

Figure 69

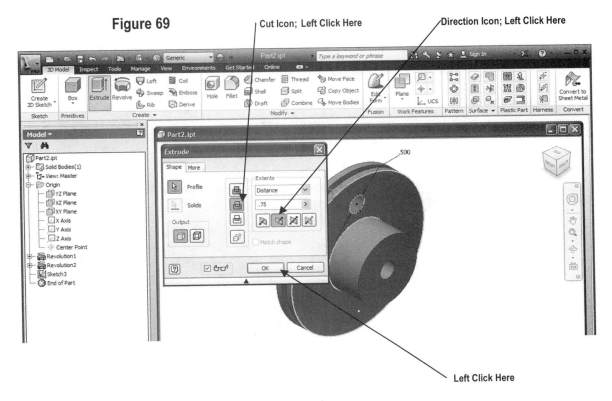

Cut Icon; Left Click Here

Direction Icon; Left Click Here

Left Click Here

76. Your screen should look similar to Figure 70.

Figure 70

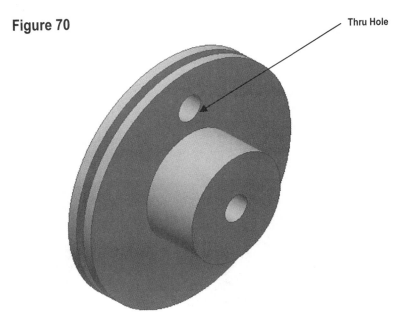

Thru Hole

Create a series of holes using the Circular Pattern command

77.   Move the cursor to the upper right portion of the screen and left click on the **Circular** icon. Earlier versions of Inventor have the Circular Pattern icon located at the middle left portion of the screen with text on the icon. If this is the case, you may have to scroll down to see the command. The Circular Pattern dialog box will appear as shown in Figure 71.

Figure 71

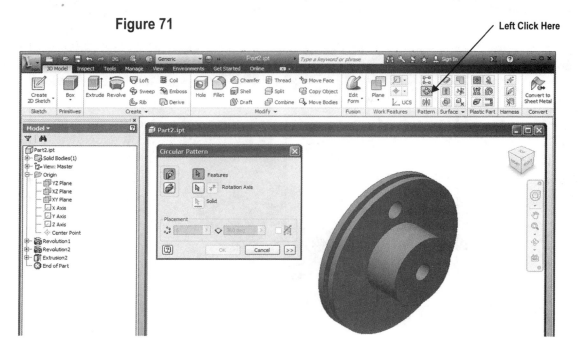

78.   Move the cursor to the center of the hole causing red dashed lines to appear and left click once. *The part must be displayed in Home/Isometric view for Inventor to find the hole as shown in Figure 72.*

Figure 72

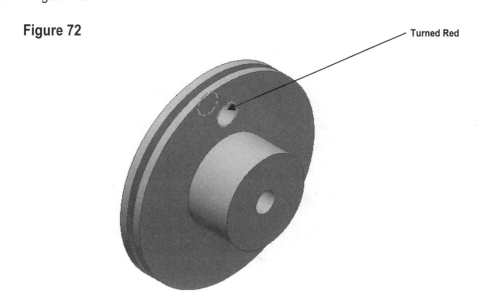

79.   Left click on **Rotation Axis** in the dialog box as shown in Figure 73.

Figure 73

Left Click Here

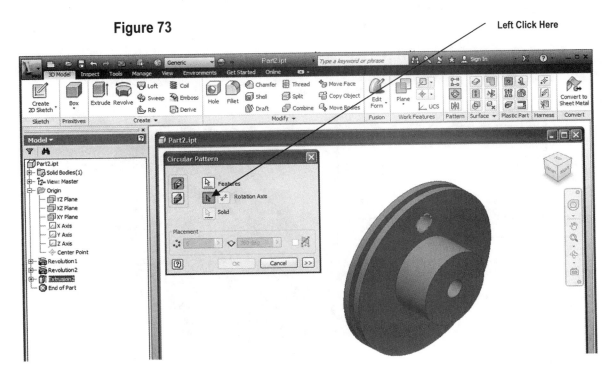

80.   Move the cursor to the edge of the part. The edges will turn red as shown in Figure 74.

Figure 74

Edges Turned Red

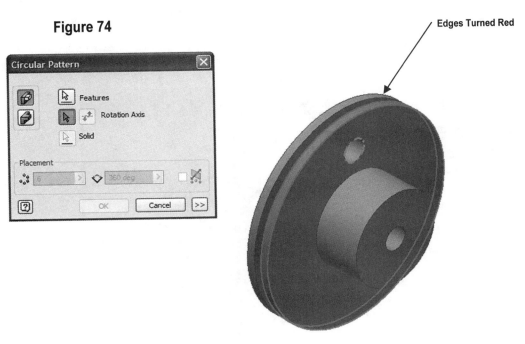

81. After the edges turn red, left click once. Inventor will provide a preview of the hole pattern as shown in Figure 75.

Figure 75

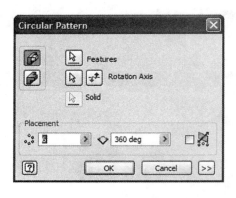

82. There are options in the Circular Pattern dialog box that are used for dictating the number of holes to be produced and the number of degrees betweens the holes. Verify that **6** is displayed for the number of holes. Verify that **360 deg** is displayed for the number of degrees. Left click on **OK** as shown in Figure 76.

Figure 76

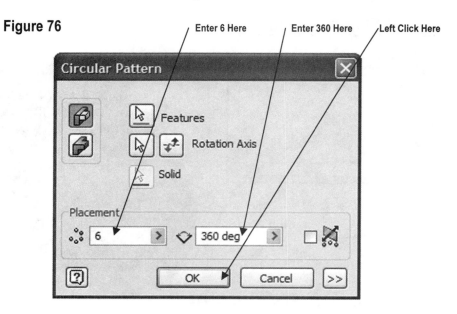

83. Your screen should look similar to Figure 77.

Figure 77

Chapter Problems

Use the Revolve and Revolve Cut Commands to complete the following.

Problem 2-1

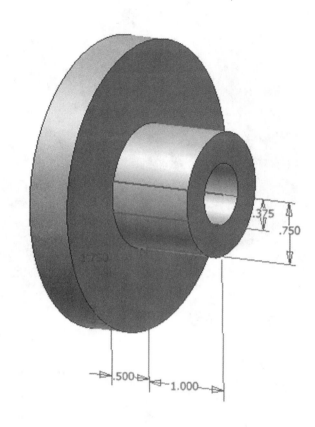

Problem 2-2

Hint: Create the solid revolve and then use the
Revolve Cut command to create the groove

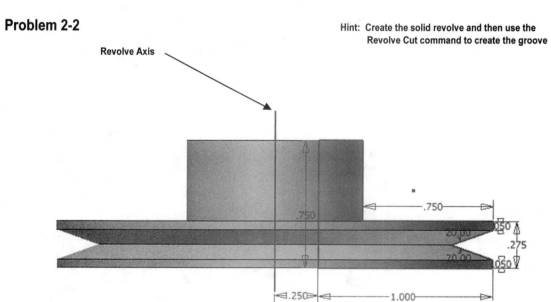

Problem 2-3

Hint: Create the solid revolve and then use the
Revolve Cut command to create the groove
8 Holes Equally Spaced

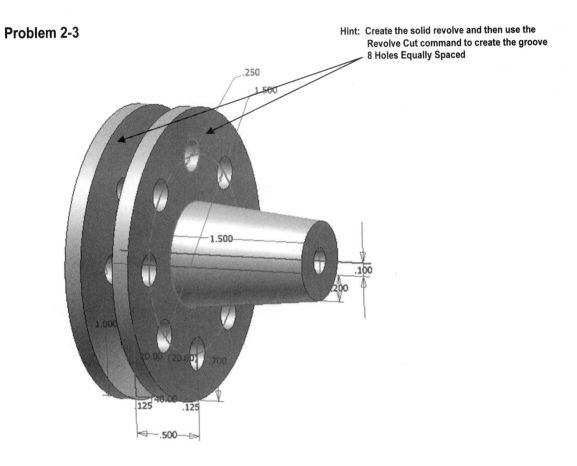

Problem 2-4

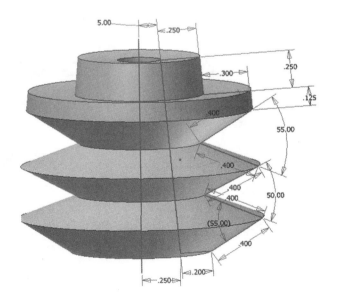

Problem 2-5

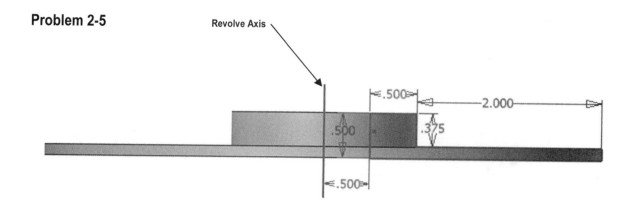

Problem 2-6

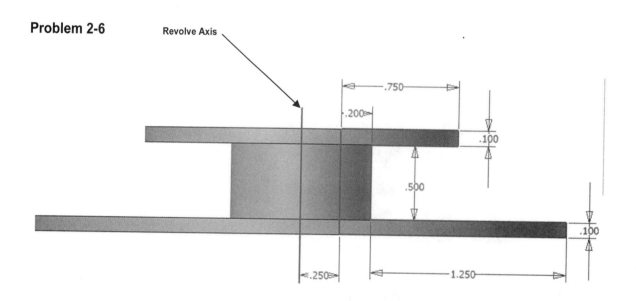

Problem 2-7

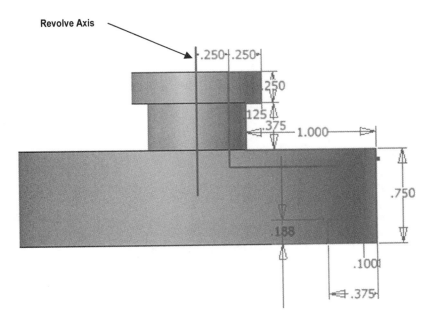

Problem 2-8

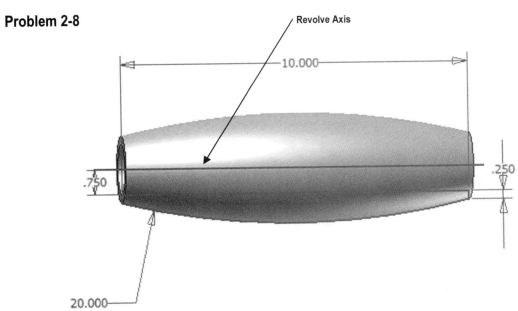

Problem 2-9

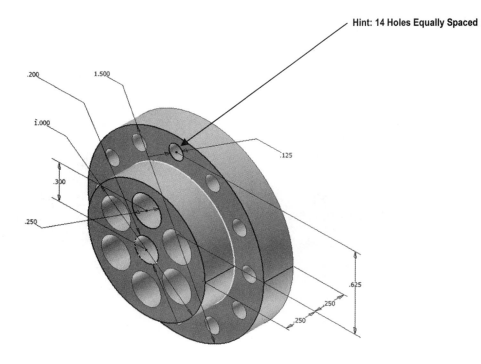

Hint: 14 Holes Equally Spaced

Problem 2-10

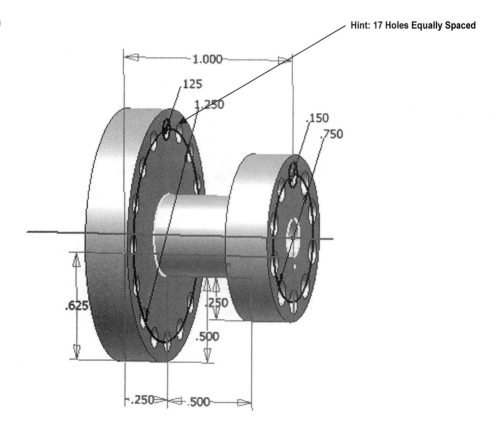

Hint: 17 Holes Equally Spaced

Problem 2-11

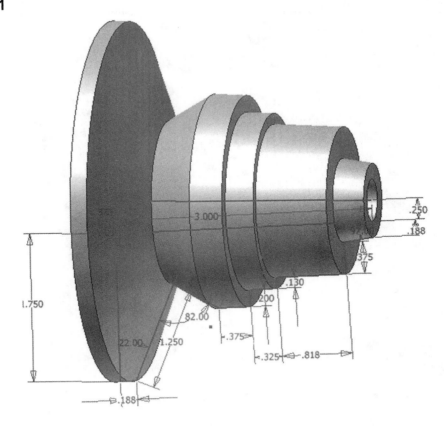

Problem 2-12

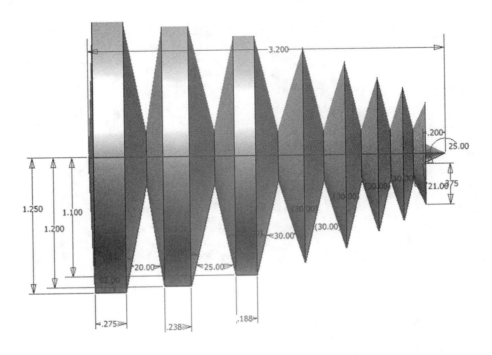

Notes:

CHAPTER 3

Learning to
Create a
Detail Drawing

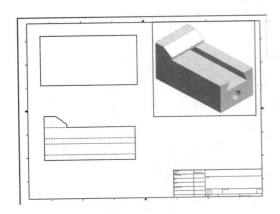

Objectives:

1. Create a simple sketch using the Sketch Panel
2. Extrude a sketch into a solid using the Model/Part Features Panel
3. Create an Orthographic view using the Place Views/Drawing Views Panel Extrude a sketch in the Model/Part Features
4. Edit the appearance of a Solid Model using the Edit Views command

Chapter 3 includes instruction on how to design the parts shown

1. Start Autodesk Inventor 2014 by referring to "Chapter 1 Getting Started".

2. After Autodesk Inventor 2014 is running, begin a new sketch.

3. Complete the drawing shown in Figure 1.

Figure 1

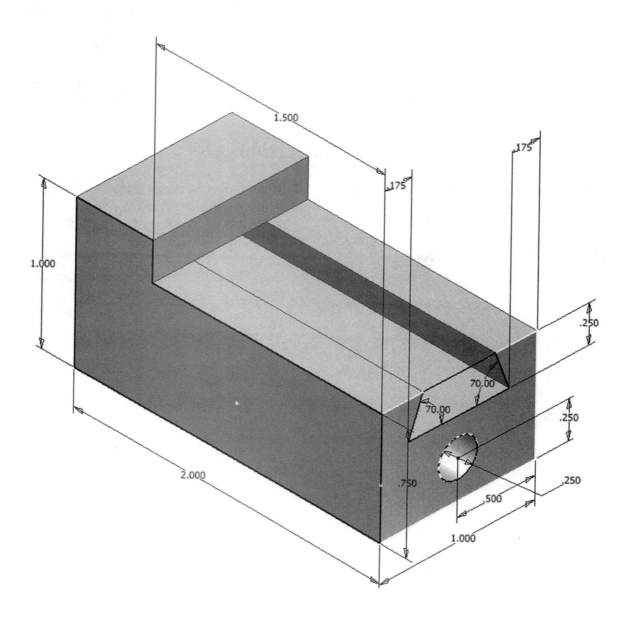

4. Move the cursor to the upper middle portion of the screen and left click on **Chamfer.** The Chamfer dialog box will appear. Left click on the drop down arrow to fully expand the dialog box as shown in Figure 2.

Figure 2

Left Click Here

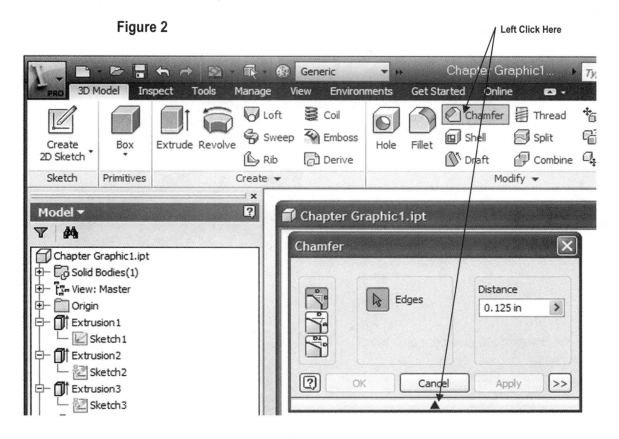

5. After selecting **Chamfer,** left click on the "Two Distance Chamfer" icon. Left click on the **Edge** icon as shown in Figure 3.

Figure 3

Left Click Here

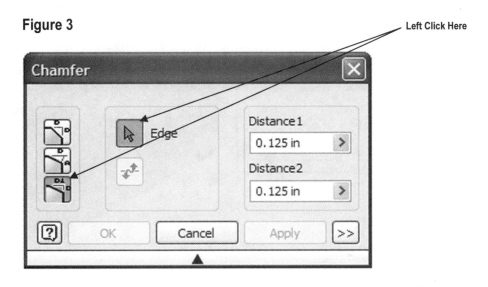

6. Move the cursor to the front upper corner. A red line will appear as shown in Figure 4.

Figure 4

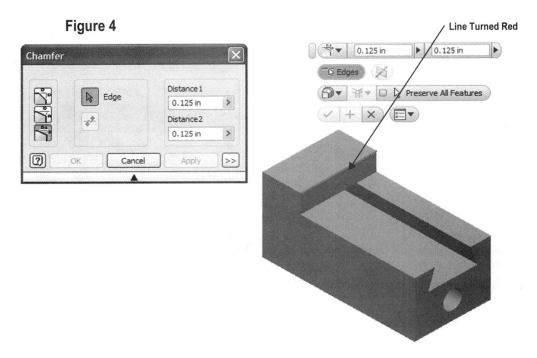

Line Turned Red

7. Inventor will provide a preview of the anticipated chamfer as shown in Figure 5.

Figure 5

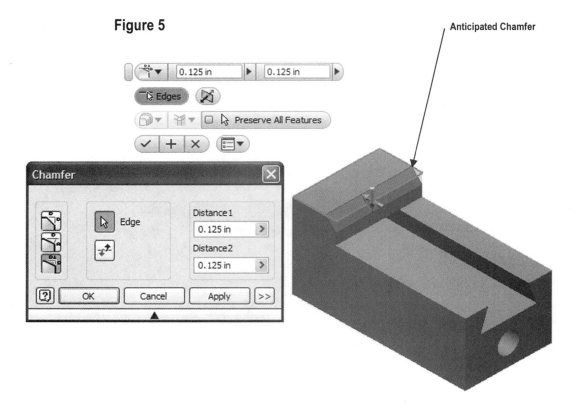

Anticipated Chamfer

8. Move the cursor to Distance 1 in the dialog box and highlight the text. Enter **.25** in the dialog box. Inventor will provide a preview of the chamfer as shown in Figure 6.

Figure 6

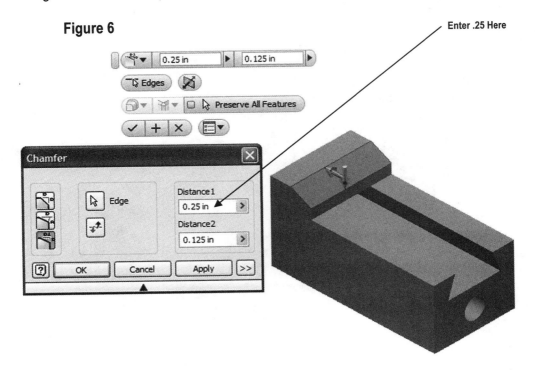

9. Move the cursor to Distance 2 in the dialog box and highlight the text. Enter **.1875** in the dialog box. Inventor will provide a preview of the chamfer as shown in Figure 7.

Figure 7

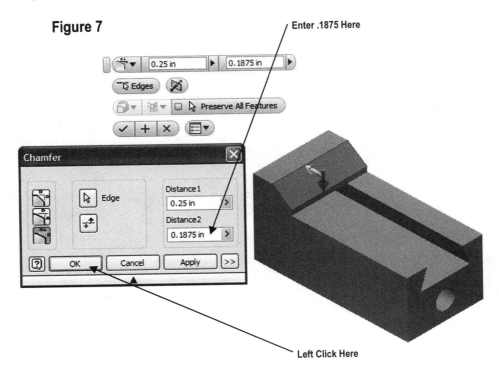

10. Left click on **OK**. Your screen should look similar to Figure 8.

Figure 8

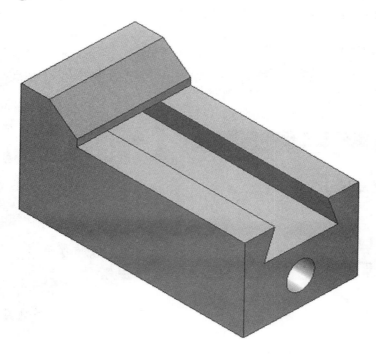

11 Save the part file for easy retrieval to be used in the following section. Do not close the part file.

12. After the part file has been saved, move the cursor to the upper left portion of the screen and left click on the "New" icon as shown in Figure 9.

Figure 9

Left Click Here

Create an Orthographic view using the Drawing Views Panel

13. The Create New File dialog box will appear. Left click on the **English** folder. Left click on the **ANSI (in).idw.** Left click on **Create** as shown in Figure 10.

Figure 10

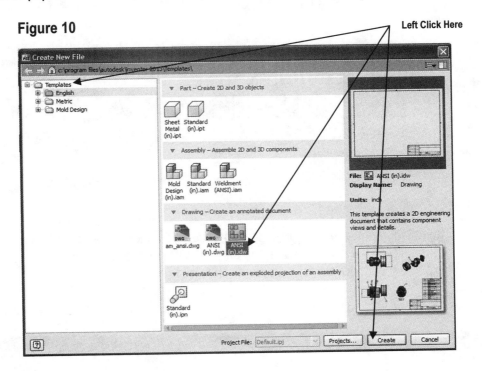

14. Your screen should look similar to Figure 11.

Figure 11

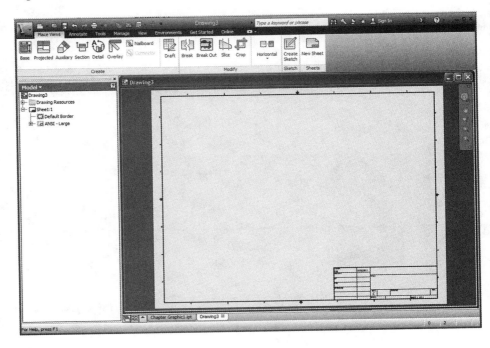

15. Inventor is now in the Place Views Panel. Notice the commands at the top of the screen are now different. The width of the screen has been reduced to add instructional clarification.

16. Move the cursor to the upper left portion of the screen and left click on **Base** as shown in Figure 12.

Figure 12

17. The drawing of the wedge block should appear attached to the cursor. Move the cursor around to verify it is attached. If the part does not appear attached to the cursor, use the "Explore" icon to locate the part file as shown in Figure 13.

Figure 13

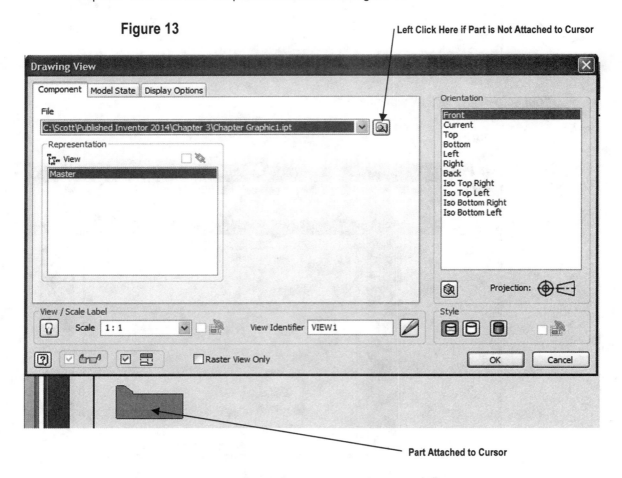

18. Different views can be selected for the front, top, and side views. Select the desired view from the Orientation selection box as shown in Figure 14. To understand how the orientation selection works, left click on **Top** or **Left** to have the top view or left view as the front (base) view.

Figure 14

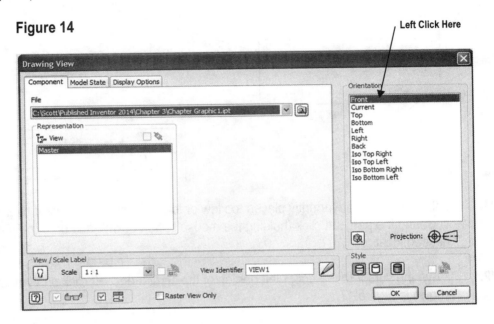

19. Select the **Front** view for the base view. Left click on the **Scale** drop down box and set the drawing scale to **4:1**. The size of the wedge block will become larger as shown in Figure 15.

Figure 15

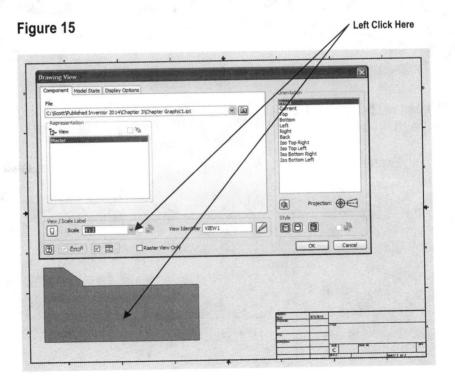

20. Place the part just above the title block that is in the lower right corner of the screen and left click once. This will place the part as shown in Figure 16.

Figure 16

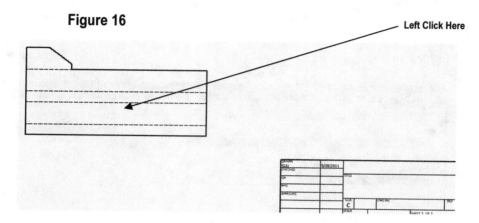

21. If the part was inadvertently placed too low or too high, move the cursor over the dots that surround the part, left click (holding the mouse button down), and drag the part to the desired location.

22. Move the cursor to the upper left portion of the screen and left click on **Projected** as shown in Figure 17.

Figure 17

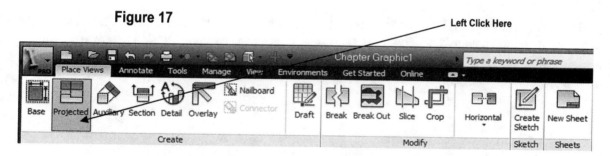

23. The part will be attached to the cursor. Move the cursor upward and left click as shown in Figure 18.

Figure 18

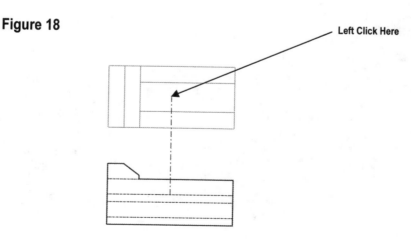

24.   Notice the black lines around the top view as shown in Figure 19. This indicates that the view has been placed.

Figure 19

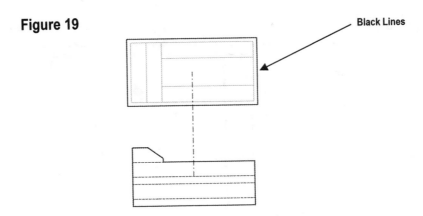

Black Lines

Create a Solid Model using the Edit Views command

25.   Move the cursor over to the upper right corner of the page and left click once as shown in Figure 20.

Figure 20

Left Click Here

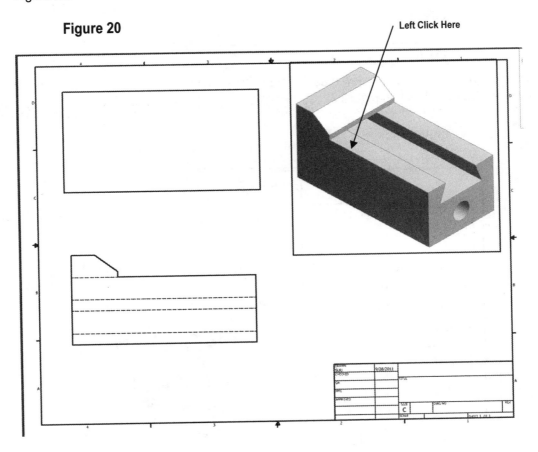

26. Move the cursor down to where the side view will be located and left click once as shown in Figure 21.

Figure 21

Left Click Here

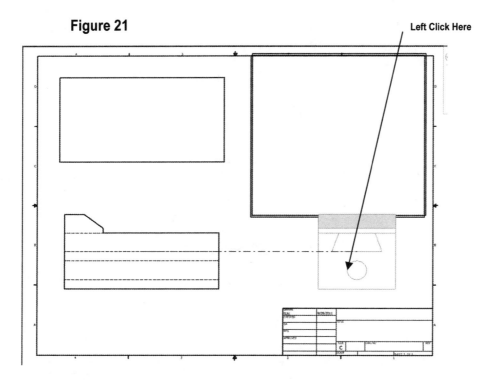

27. Right click on the last view created (side view). A pop up menu will appear. Left click on **Create** as shown in Figure 22.

Figure 22

Left Click Here

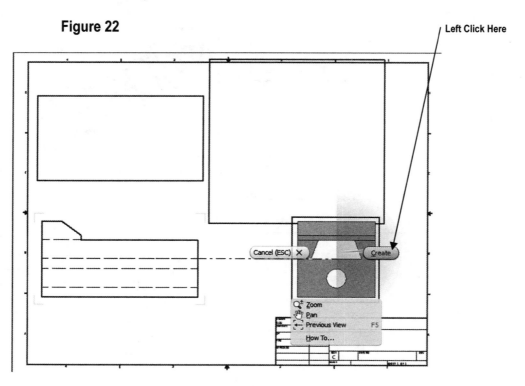

28.   Your screen should look similar to Figure 23.

Figure 23

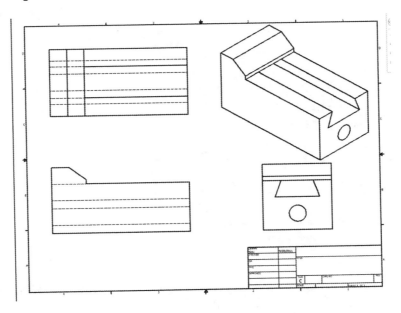

29.   Move the cursor over the isometric view in the upper right corner of the drawing. Red dots will appear as shown in Figure 24.

Figure 24

Red Dots Appear

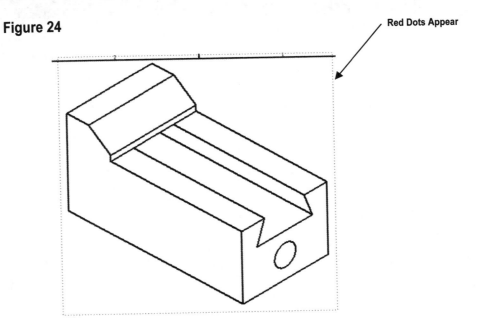

30. After the red dots appear, right click once. A pop up menu will appear. Left click on **Edit View** as shown in Figure 25.

Figure 25

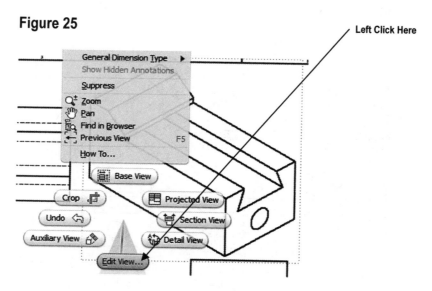

Left Click Here

31. The Drawing View dialog box will appear. Left click on the "blue barrel" to the right under Style. Left click on **OK** as shown in Figure 26.

Figure 26

Left Click Here

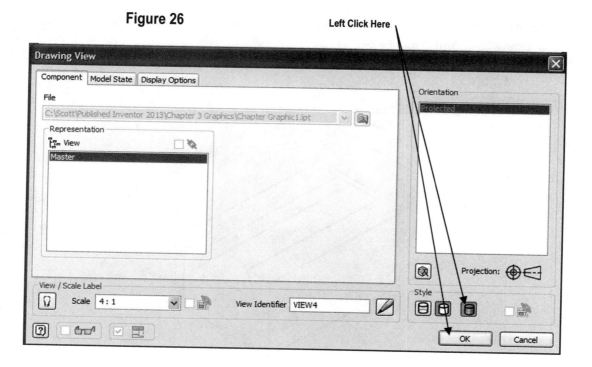

32. The isometric view will become a miniature solid model as shown in Figure 27.

Figure 27

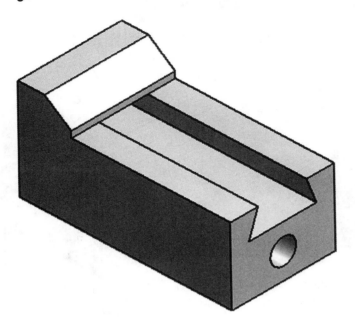

33. Your screen should look similar to Figure 28.

Figure 28

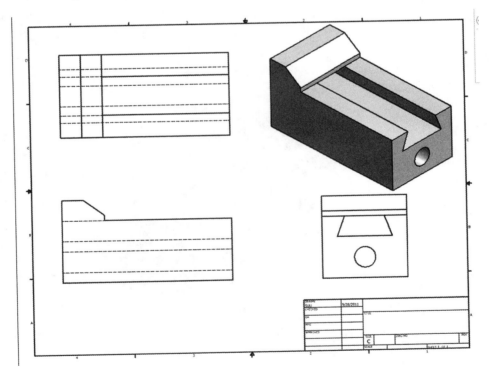

Chapter Problems

Create 3 view/multi view drawings of the following parts.

Problem 3-1

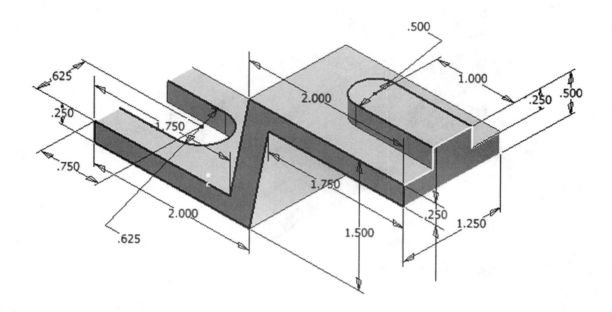

Problem 3-2

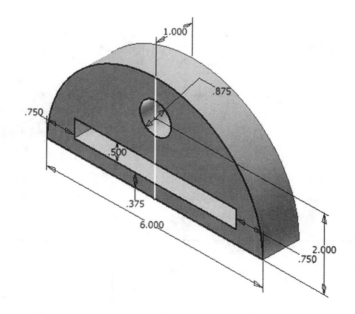

Problem 3-3

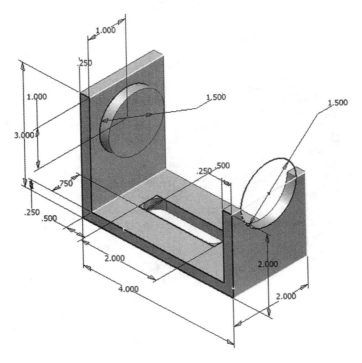

Problem 3-4

Hint: Use the 2 directional
Extrude-Cut to create
The 1.625 Diameter Hole

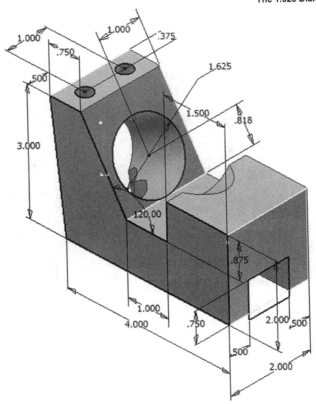

Problem 3-5

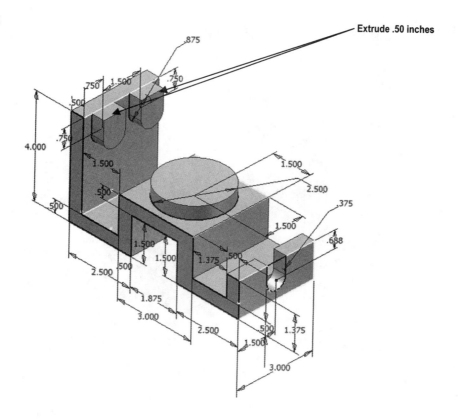

Problem 3-6

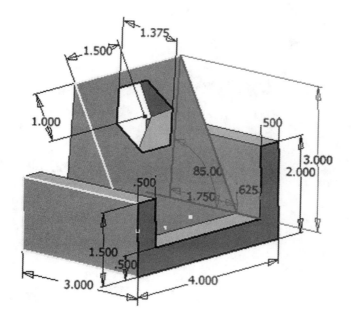

Problem 3-7

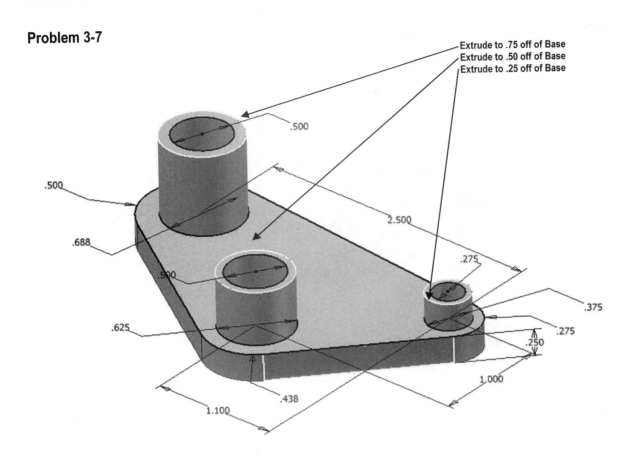

Extrude to .75 off of Base
Extrude to .50 off of Base
Extrude to .25 off of Base

Problem 3-8

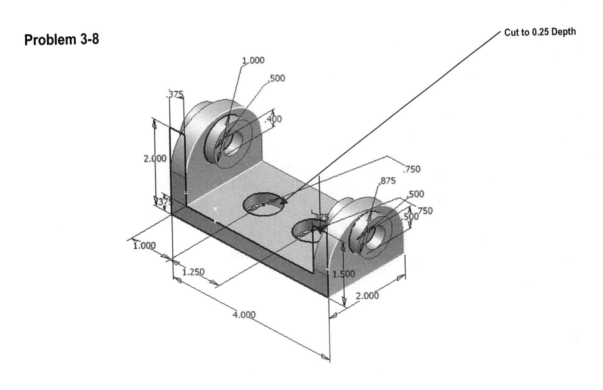

Cut to 0.25 Depth

Notes:

Advanced Detail Drawing Procedures

Objectives:

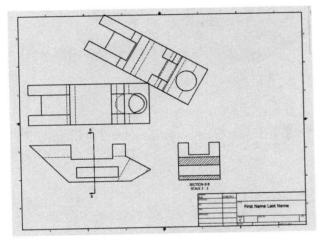

1. Create an Auxiliary View using the Place Views/Drawing Views Panel
2. Create a Section View using the Place Views /Drawing Views Panel
3. Dimension views using the Annotation/ Drawing Annotation Panel
4. Create Text using the Annotation/ Drawing Annotation Panel

Chapter 4 includes instruction on how to create the drawings shown.

1.   Start Autodesk Inventor 2014 by referring to "Chapter 1 Getting Started".

2.   After Autodesk Inventor 2014 is running, complete the following part as shown in Figure 1.

Figure 1

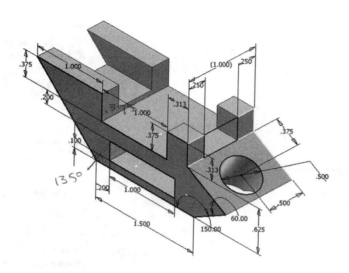

3.   Once the part is complete project the part into a 3 view drawing as discussed in Chapter 3 using a 3:1 scale as shown in Figure 2.

Figure 2

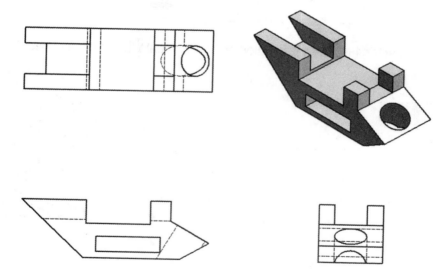

4.   Start by moving the views closer together to provide additional room on the drawing. Move the cursor over the top view causing dots to appear around the view. After the dots appear, left click on the dots (holding the left mouse button down) and drag the view down closer to the front (base) view as shown in Figure 3.

Figure 3

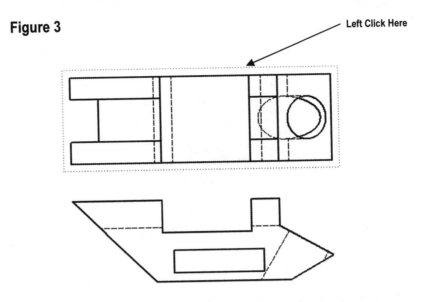

Left Click Here

5.   Move the side view closer to the front (base) view.   Start by moving the cursor over the side view causing dots to appear around the view. After the dots appear, left click on the dots (hold the left mouse button down) and drag the view closer to the front (base) view as shown in Figure 4.

Figure 4

Left Click Here

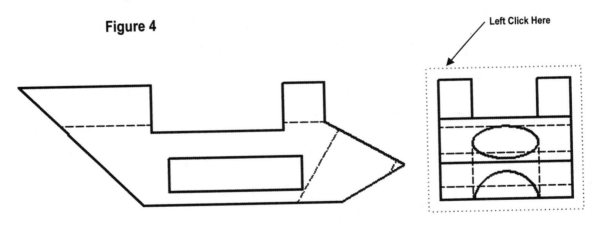

6. You will need to delete the isometric view that was created in Chapter 3. Move the cursor near the isometric view causing red dots to appear. Right click. A pop up menu will appear. Left click on **Delete** as shown in Figure 5.

Figure 5

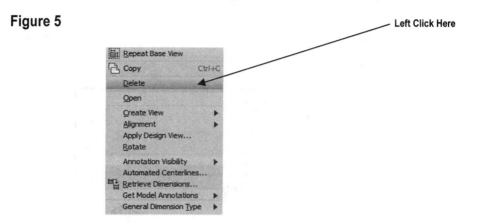

Left Click Here

7. The Delete dialog box will appear. Left click on **OK** as shown in Figure 6.

Figure 6

Left Click Here

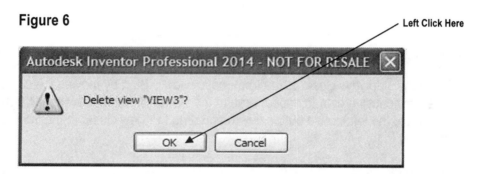

8. There will now be more room to work. Your screen should look similar to Figure 7.

Figure 7

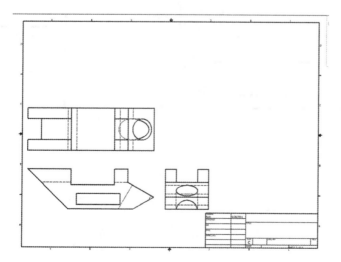

9. To provide more space on the drawing, the drawing view scale will have to be reduced. Right click on the front (base) view. A pop up menu will appear. Left click on **Edit View** as shown in Figure 8.

Figure 8

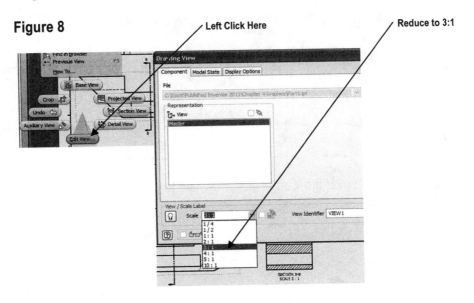

Create an Auxiliary View using the Drawing Views Panel

10. Move the cursor to the upper left portion of the screen and left click on **Auxiliary View** as shown in Figure 9.

Figure 9

11. Move the cursor to the front (base) view causing red dots to appear around the view as shown in Figure 10. Left click once.

Figure 10

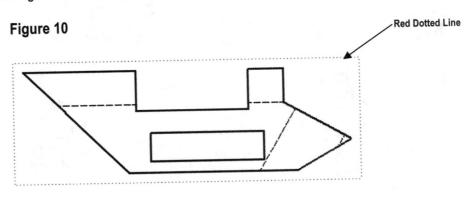

12. The Auxiliary View dialog box will appear as shown in Figure 11.

Figure 11

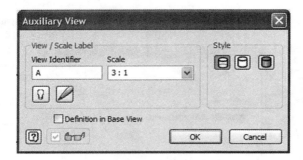

13. Move the cursor over the wedge line causing it to turn red. Left click as shown in Figure 12.

Figure 12

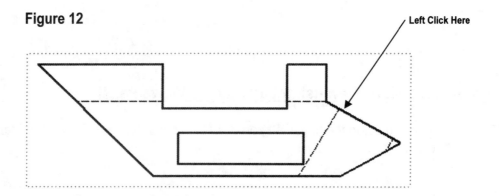

Left Click Here

14. Inventor will create an auxiliary view from the selected surface. The view will be attached to the cursor as shown in Figure 13.

Figure 13

Auxiliary View

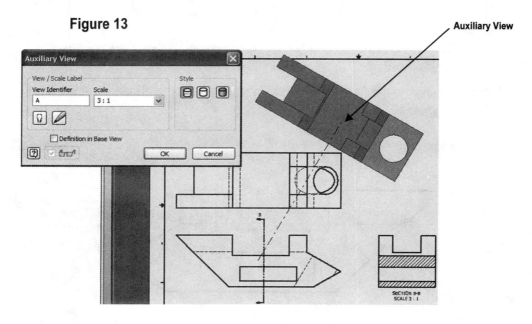

15. Move the cursor towards the upper right and left click. The Auxiliary View dialog box will close as shown in Figure 14.

Figure 14

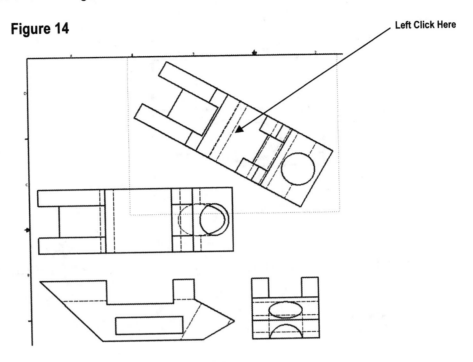

16. Your screen should look similar to Figure 15.

Figure 15

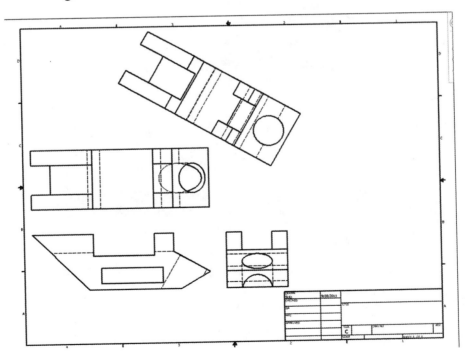

17.   Move the cursor to the side view causing red dots to appear as shown in Figure 16.

Figure 16

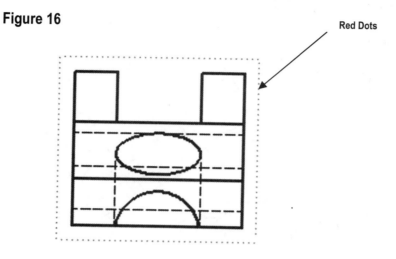

18.   Right click on the view. A pop up menu will appear. Left click on **Delete** as shown in Figure 17.

Figure 17

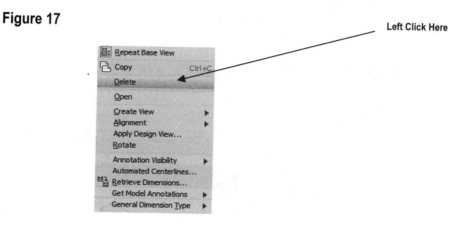

19.   A Delete dialog box will appear. Left click on **OK** as shown in Figure 18.

Figure 18

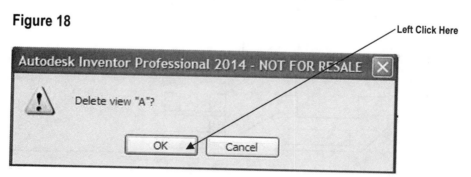

Create a Section View using the Drawing Views Panel

20. Move the cursor to the upper left portion of the screen and left click on **Section View** as shown in Figure 19.

Figure 19

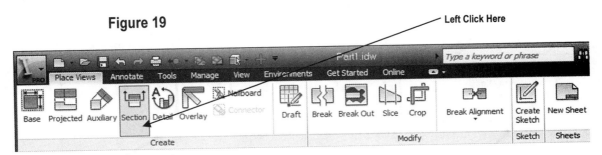

21. Move the cursor over the front view causing red dots to appear around the view as shown in Figure 20.

Figure 20

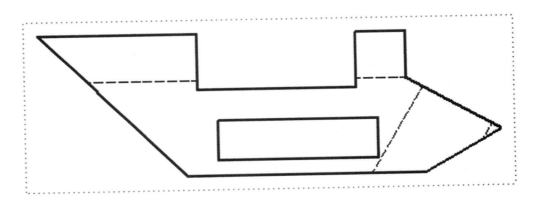

22. Left click on the view causing the red dots to turn into a solid red line as shown in Figure 21.

Figure 21

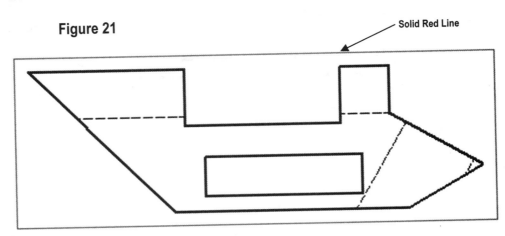

23.   Move the cursor around outside the red line and wait for the dotted line to appear as shown in Figure 22. It may take a few seconds before the line appears.

Figure 22

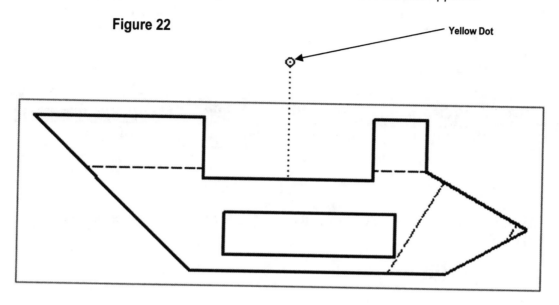

24.   Left click on the yellow dot, move the line down, and left click as shown in Figure 23.

Figure 23

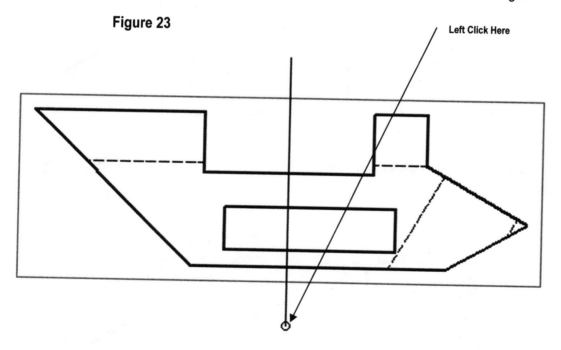

25.   Right click in the same location. A pop up menu will appear. Left click on **Continue** as shown in Figure 24.

Figure 24

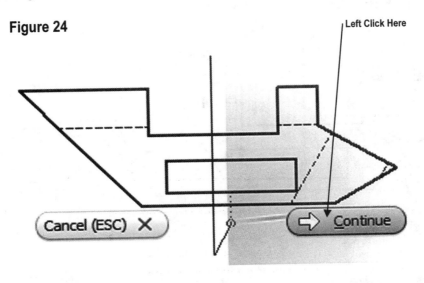

26.   The Section View dialog box will appear as shown in Figure 25. The section view will be attached to the cursor. Move the cursor to the right where the side view was located and left click once. The Section View dialog box will close.

Figure 25

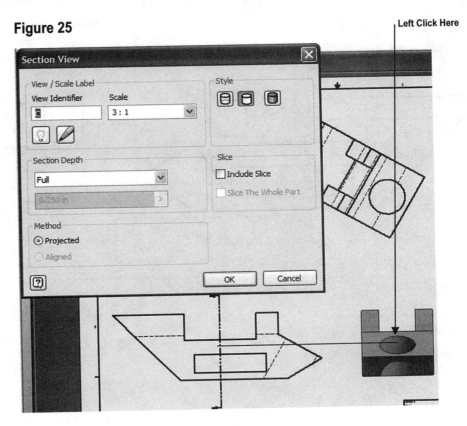

4-11

27.   Inventor will create a section view to the right as shown in Figure 26.

Figure 26

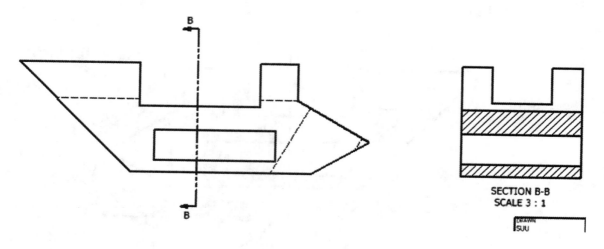

28.   Inventor will create a section view that represents wherever the cutting plane line cuts through the part. Move the cursor over the center of the cutting plane line causing it to turn red as shown in Figure 27.

Figure 27

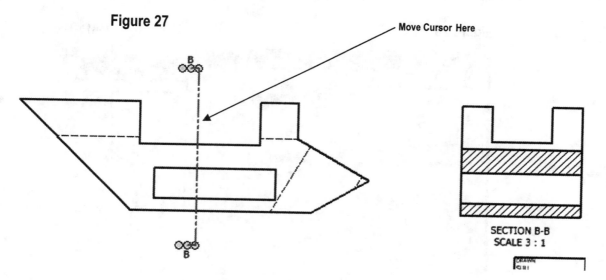

29. Once the line becomes highlighted (turns red) left click (holding the left mouse button down) and drag the cursor to the right. The cutting plane line will become a normal looking line while attached to the cursor as shown in Figure 28.

Figure 28

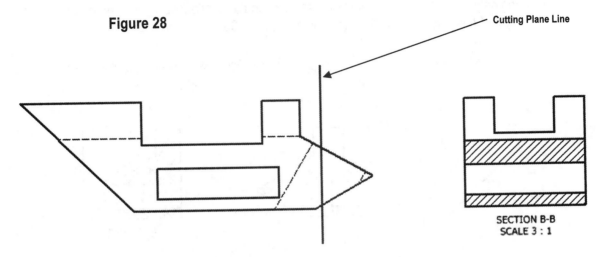

30. Once the cutting plane line has been moved to a new location release the left mouse button. The side view now reflects the new location of the cutting plane line as shown in Figure 29.

Figure 29

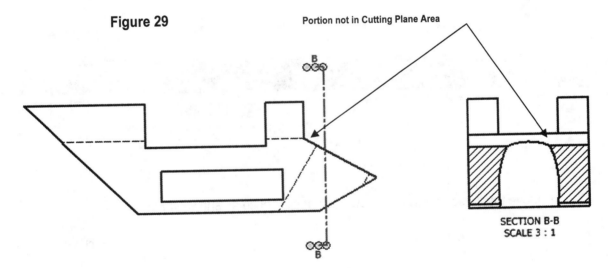

Create a broken view using the Break command

31. Left click (holding the left mouse button down) on the cutting plane line and move it back to its original location. Notice that the cross hatch in the section view will update to reflect the location of the cutting plane line as shown in Figure 30.

Figure 30

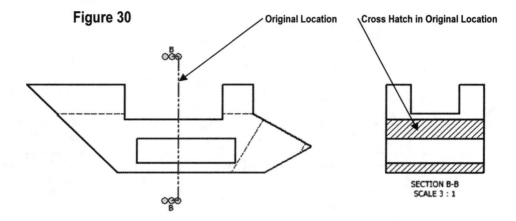

32. Move the cursor to the middle left portion of the screen and left click on **Break** as shown in Figure 31.

Figure 31

33. Move the cursor to the front view and left click once as shown in Figure 32.

Figure 32

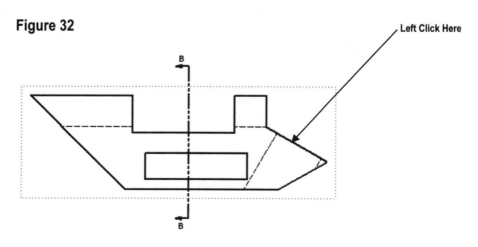

34. The Break Dialog box will appear.

Figure 33

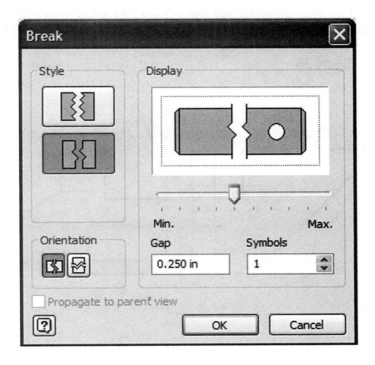

35. Left click on the part causing a red box to appear as shown in Figure 34.

Figure 34

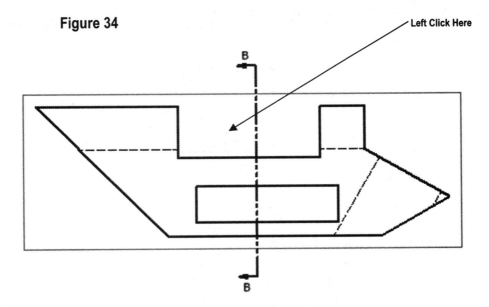

36. Move the cursor to the left side of the cutting plane line and left click once. Move the cursor to the left. Another line will appear next to the first line. These two lines represent the size of the gap that Inventor will create in the part. A third line will be attached to the cursor. This line represents how much of the part will be removed from the view. Move the cursor to the far left portion of the part and left click as shown in Figure 35.

Figure 35

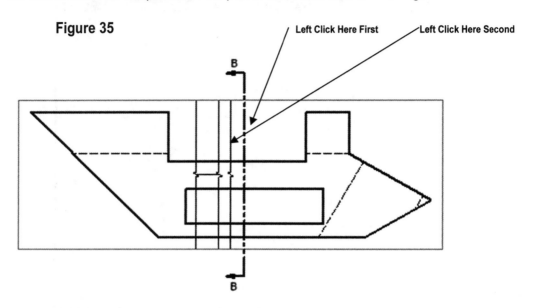

37. Inventor will remove sections from both the front and top views as shown in Figure 36.

Figure 36

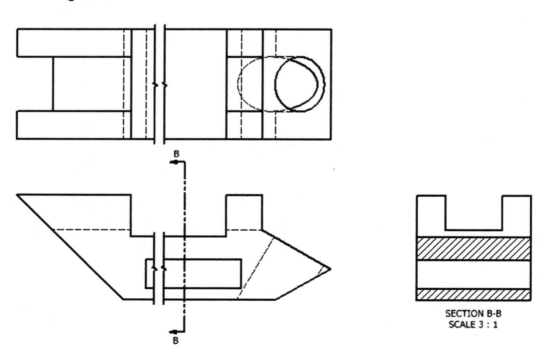

SECTION B-B
SCALE 3 : 1

38. Move the cursor to the upper left portion of the screen and left click on **Undo** as shown in Figure 37.

Figure 37

Dimension views using the Drawing Annotation Panel

39. Move the cursor to the upper left portion of the screen and left click on the **Annotation** tab. Left click on **Dimension** as shown in Figure 38.

Figure 38

40. Move the cursor over the top horizontal line causing it to turn red and left click once. Then move the cursor over the bottom horizontal line causing it to turn red and left click once as shown in Figure 39. The dimension will be attached to the cursor.

Figure 39

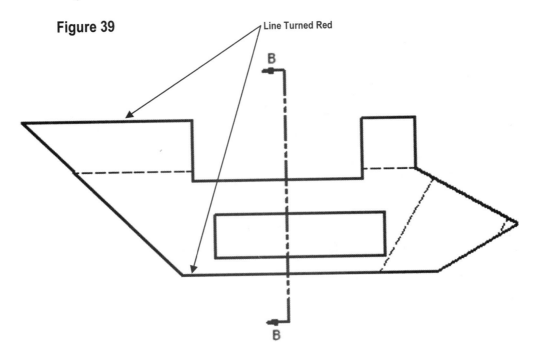

41. Move the cursor to the left and left click once. The actual dimension of the line will appear as shown in Figure 40. The Edit Dimension dialog box (not shown) will appear. Left click on **OK**.

Figure 40

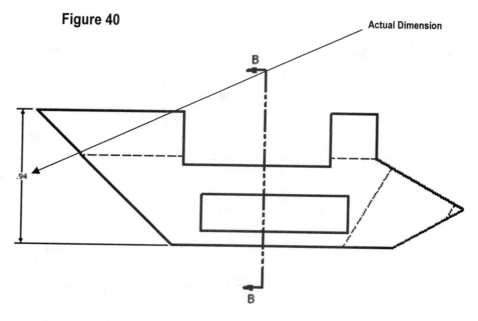

42. Finish dimensioning the part to your own satisfaction. When the part is satisfactorily dimensioned, save the file to a location where it can easily be retrieved.

43. To delete an unwanted dimension, move the cursor over the dimension. The dimension will turn red and several green dots will appear as shown in Figure 41.

Figure 41

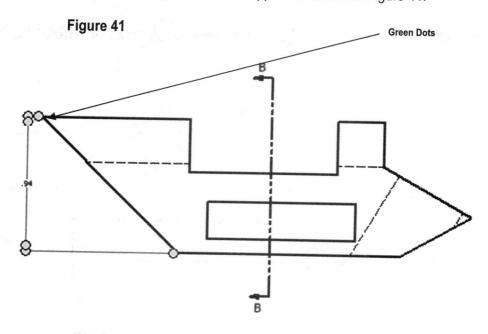

Create Text using the Drawing Annotation Panel.

44. Right click on the dimension. A pop up menu will appear. Left click on **Delete** as shown in Figure 42.

Figure 42

Left Click Here

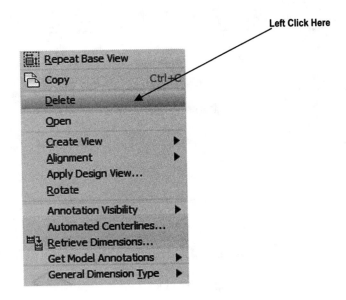

45. Move the cursor to the upper middle portion of the screen and left click on **Text** as shown in Figure 43.

Figure 43

Left Click Here

46. Move the cursor to the title block location as shown in Figure 44. Left click once when the yellow dot appears.

Figure 44

Left Click Here

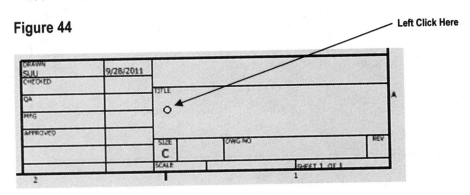

47. The Format Text dialog box will appear. Left click on the drop down box and change the text height to **.240** inches as shown in Figure 45.

Figure 45

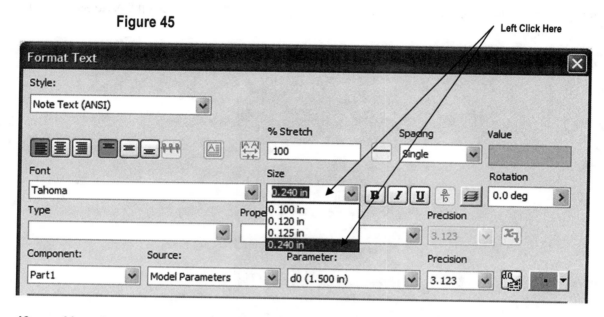

48. Move the cursor to the open area located in the lower half of the Format Text dialog box and enter your first and last name. Text will appear near the flashing cursor as shown in Figure 46.

Figure 46

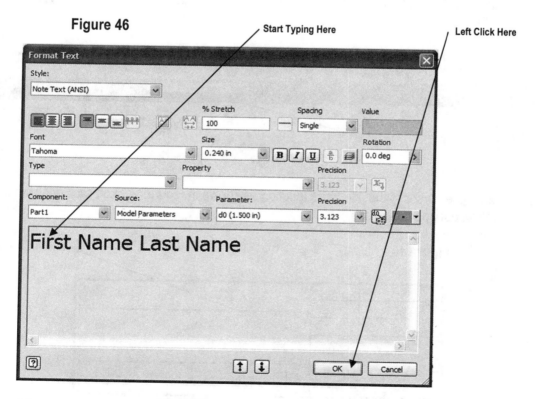

49. After text has been entered, left click on **OK** as shown in Figure 46.

50. The Format Text dialog box will close.

51. Text will appear in the title block as shown in Figure 47.

Figure 47

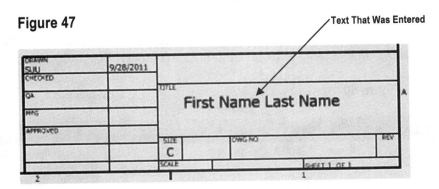

52. Right click near the text. A pop up menu will appear as shown in Figure 48.

Figure 48

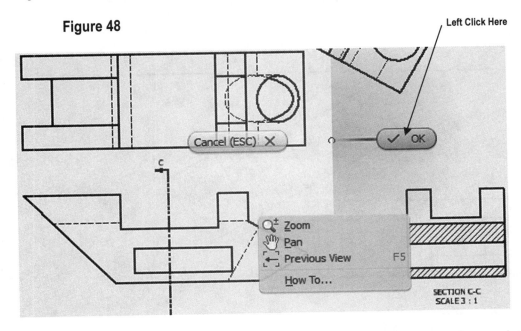

53. If the text needs to be moved, move the cursor over the text causing several green dots to appear as shown in Figure 49.

Figure 49

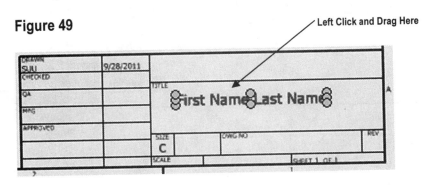

54. While the text is highlighted, left click (holding the left mouse button down) and drag the text to the desired location. After the text is in the desired location, release the left mouse button, move the cursor away from the text, and left click once.

55. Move the cursor to the upper left portion of the screen and left click on the **Place Views** tab as shown in Figure 50. This will return Inventor to the Place View options menu.

Figure 50   Left Click Here

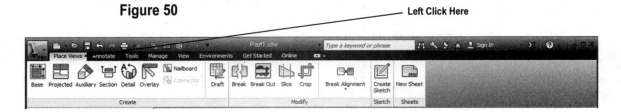

56. Your screen should look similar to Figure 51.

Figure 51

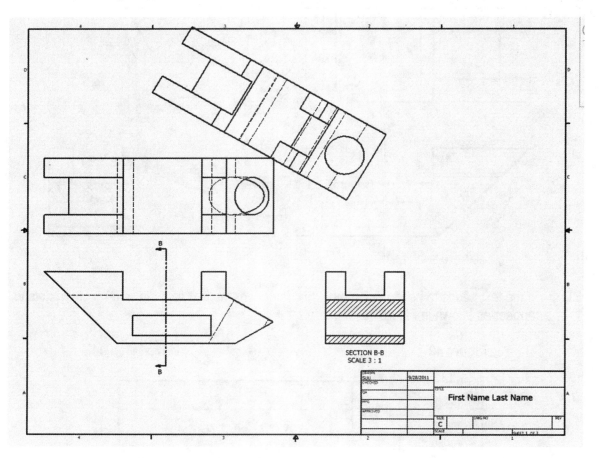

57. Before starting a new sheet of detail drawings, make sure to first save the current sheet. **Caution: Once a new sheet has been created the old sheet is not retrievable unless it has been saved. If a new sheet is created before the old sheet was saved, left click on the Undo icon located at the upper left portion of the screen as shown in Figure 52.**

Figure 52   Left Click Here if a New Sheet was Started before Saving the Existing Sheet

58. Move the cursor to the left middle portion of the screen and left click on **New Sheet** as shown in Figure 53.

Figure 53   Left Click Here

59. This will begin a new sheet for more detail drawings if necessary.

Chapter Problems

Create Section View Drawings for the following problems.

Problem 4-1

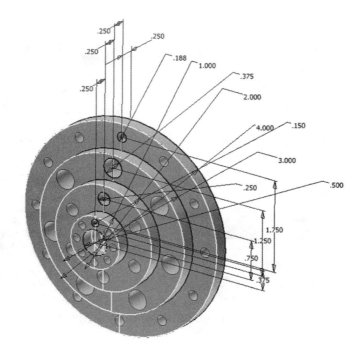

Problem 4-2

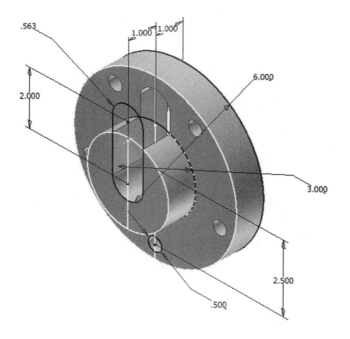

Problem 4-3   Revolve the following sketch then create a Section View

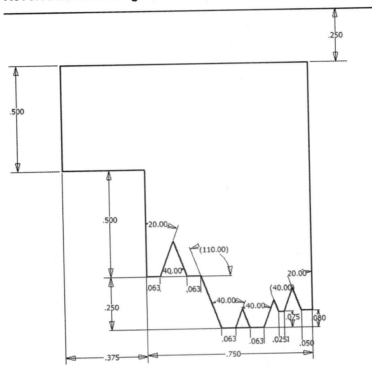

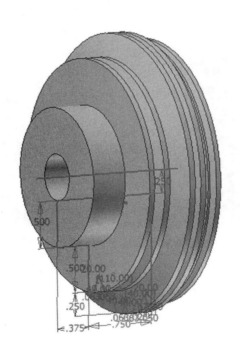

Problem 4-4   Revolve the following sketch then create a Section View

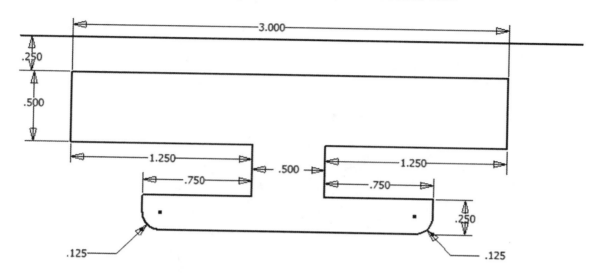

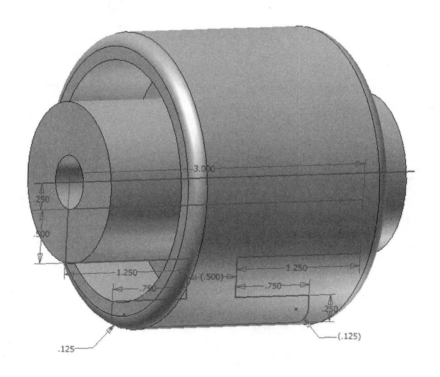

Problem 4-5   Create Section View Drawings for the following problems

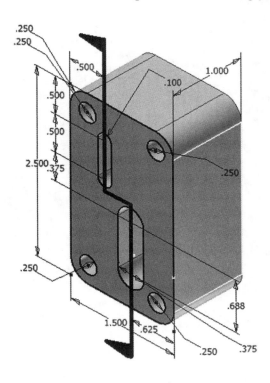

Problem 4-6

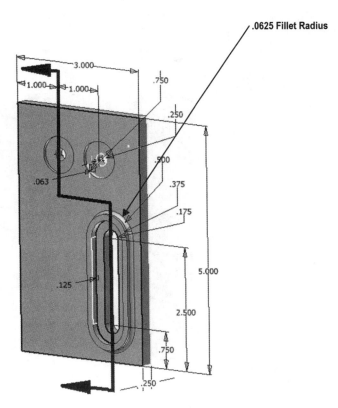

.0625 Fillet Radius

Problem 4-7

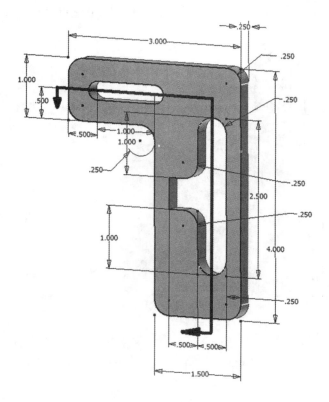

Problem 4-8

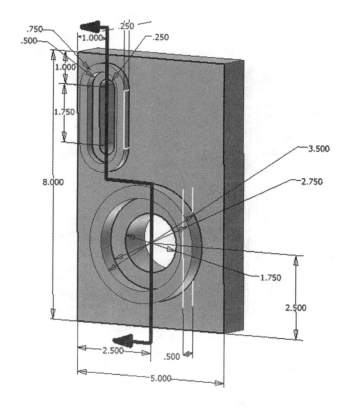

CHAPTER 5

Learning To Edit Existing Solid Models

Objectives:

1. Design a simple part
2. Learn to use the Circular Pattern Command
3. Edit the part using the Sketch Panel
4. Edit the part using the Extrude Command
5. Edit the part using the Fillet Command

Chapter 5 includes instruction on how to design and edit the part shown.

1. Start Autodesk Inventor 2014 by referring to "Chapter 1 Getting Started".

2. After Autodesk Inventor 2014 is running, begin a new sketch.

3. Create the sketch shown in Figure 1.

Figure 1

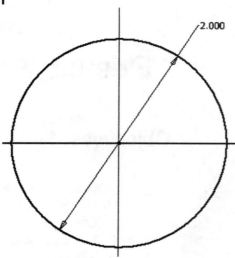

4. Change the view to Isometric/Home View as shown in Figure 2.

Figure 2

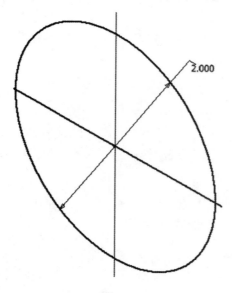

5. Extrude the sketch to a distance of .25 inches as shown in Figure 3.

6. Once a solid has been created, begin a New Sketch on the front surface as shown in Figure 3.

Figure 3

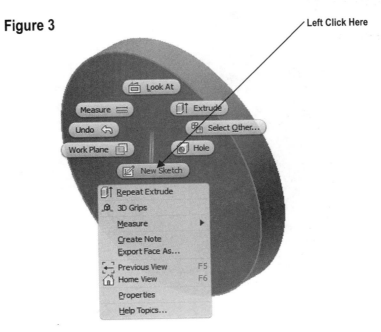

7. Complete the following sketch. Estimate the center location of the hole from the center of the part.

Figure 4

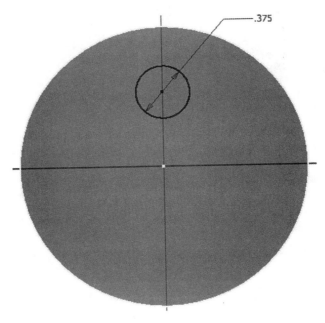

8. Exit out of the Sketch Panel. Change the view to Isometric/Home View as shown in Figure 5.

Figure 5

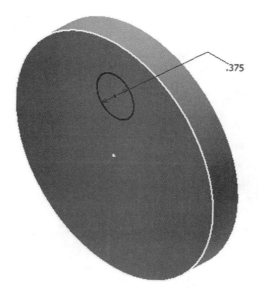

9. Use the Extrude-Cut command to cut a hole in the part as shown in Figure 6.

Figure 6

10. Use the Circular Pattern command to create 3 holes in the part as shown in Figure 7.

Figure 7

11. Use the Fillet command to create a fillet with .0625 radius as shown in Figure 8.

Figure 8

Edit the part using the Sketch Panel

12. If for some reason a change needs to be made to this part, it can be accomplished by editing either a sketch or a feature located in the Part Tree at the upper left corner of the screen as shown in Figure 9.

Figure 9

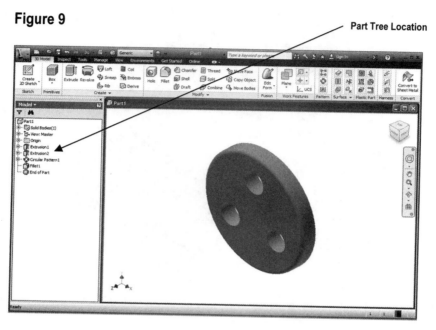

13. A close-up of the Part Tree is shown in Figure 10. Left click on each of the "plus" signs in the part tree. The tree will expand showing more details for part construction.

Figure 10

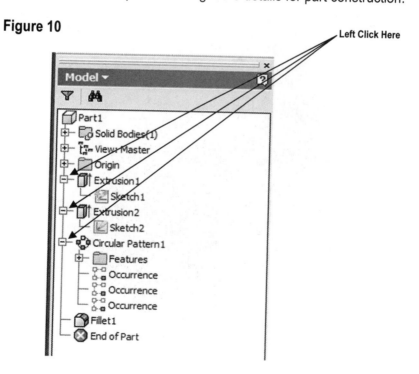

14. If a change needs to be made to any portion of the part that was constructed using a Sketch1, the change can be made here.

15. Move the cursor over Sketch1. A red box will appear around the text "Sketch1" as shown in Figure 11.

Figure 11

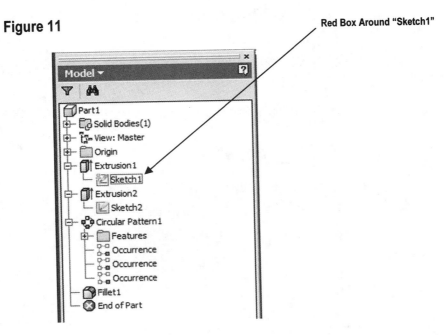

Red Box Around "Sketch1"

16. The original sketch will also appear as shown in Figure 12.

Figure 12

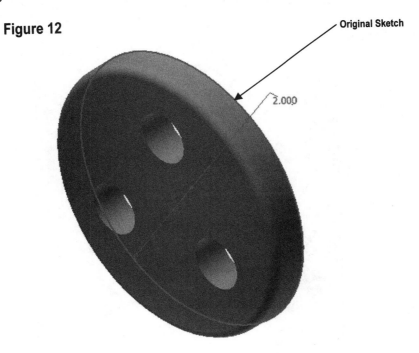

Original Sketch

2.000

17. Right click on **Sketch1**. The text "Sketch1" will become highlighted. A pop up menu will appear. Left click on **Edit Sketch** as shown in Figure 13.

Figure 13

Left Click Here

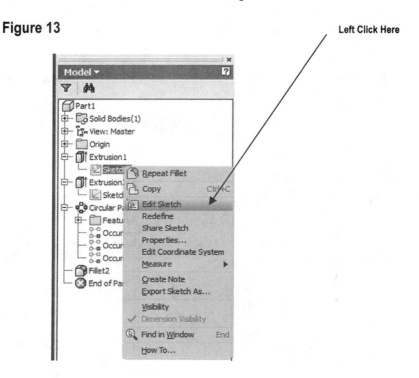

18. The original sketch will appear as shown in Figure 14.

Figure 14

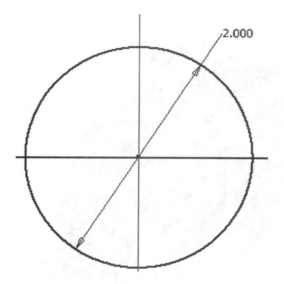

19. If Inventor rotated the part to provide a perpendicular view, then skip to Number 22. If the sketch is not a perpendicular view, move the cursor to the upper right portion of the screen and left click on the "Face View/Look At" icon as shown in Figure 15.

Figure 15

20. Move the cursor over to the part tree and left click on the "plus" sign to the left of Origin. The part tree will expand displaying all three work planes. Move the cursor over the text "XY Plane". A red box will appear around XY Plane and the sketch itself. After the red box appears, left click once on the **XY Plane** as shown in Figure 16.

Figure 16

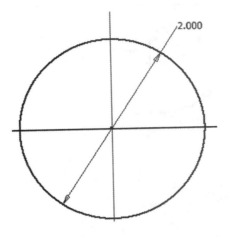

21. Inventor will provide a perpendicular view of the sketch similar to when the sketch was first constructed. Your screen should look similar to Figure 17.

Figure 17

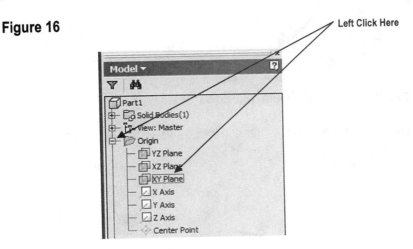

22. Start by modifying the diameter of the part. First, double click on the overall dimension. The Edit Dimension dialog box will appear as shown in Figure 18.

Figure 18

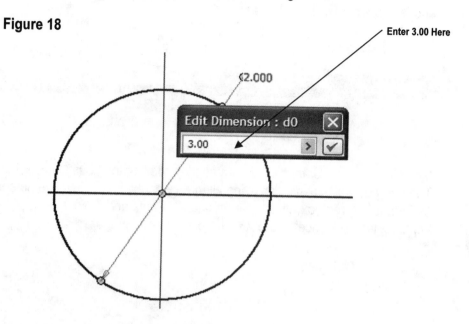

23. Enter **3.00** as shown in Figure 18. Press **Enter** on the keyboard.

24. The diameter of the part will increase to 3.00 as shown in Figure 19.

Figure 19

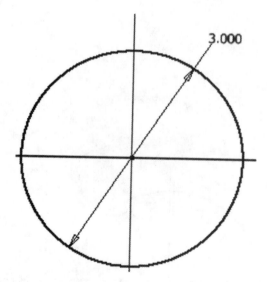

25. Move the cursor to the upper left portion of the screen and left click on the **Manage** tab. Left click on the drop down arrow below Update. A drop down menu will appear. Left click on **Update** as shown in Figure 20.

Figure 20

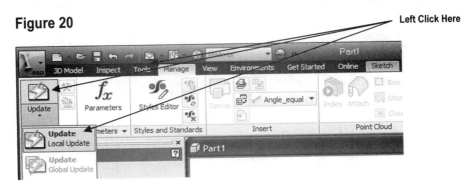

26. Inventor will automatically update the part as shown in Figure 21. The part will be updated without the need to repeat any of the steps that created the original part.

Figure 21

Edit the part using the Extrude command

27.   Move the cursor over the text "Extrusion1". A red box will appear around the text. After the red box appears, right click once on **Extrusion1** as shown in Figure 22.

Figure 22

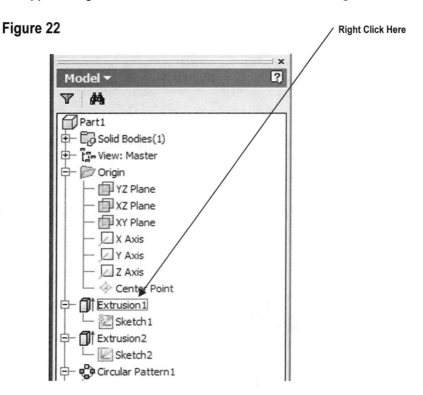

28.   A pop up menu will appear. Left click on **Edit Feature** as shown in Figure 23.

Figure 23

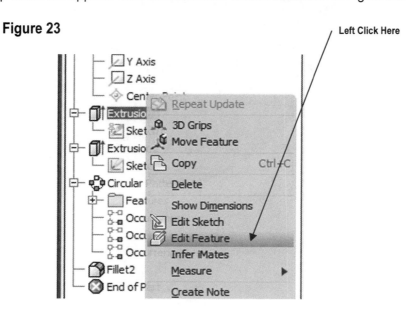

29. The Extrusion dialog box will appear. Enter **.500** for the extrusion distance and left click on **OK** as shown in Figure 24.

Figure 24

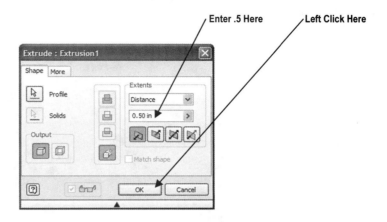

30. Move the cursor to the upper left portion of the screen and left click on the **Manage** tab. Left click on the drop down arrow below Update. A drop down menu will appear. Left click on **Update** as shown in Figure 25.

Figure 25

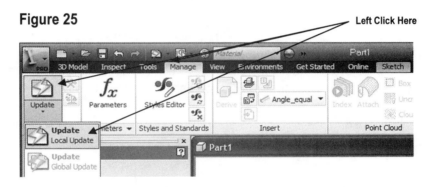

31. Inventor will automatically update the part. Notice that the holes are no longer thru holes as shown in Figure 26.

Figure 26

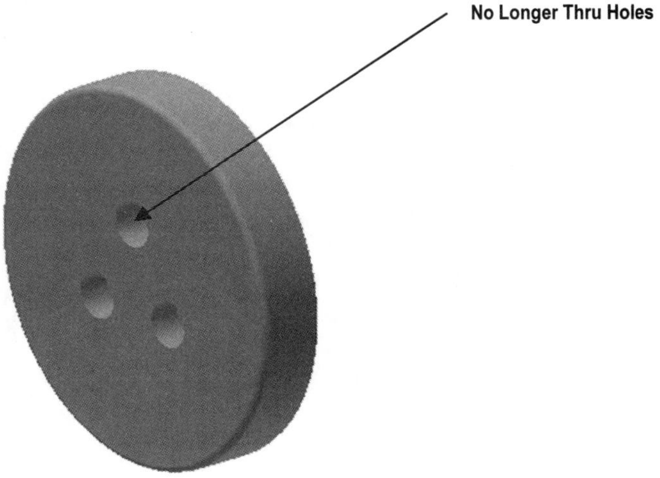

No Longer Thru Holes

32. Rotate the view around close to perpendicular using "Face View/Look At" command to see that the holes are no longer thru as shown in Figure 27.

Figure 27

33. Move the cursor over the text "Sketch2". A red box will appear around the text. After the red box appears, right click once on **Sketch2** as shown in Figure 28.

Figure 28

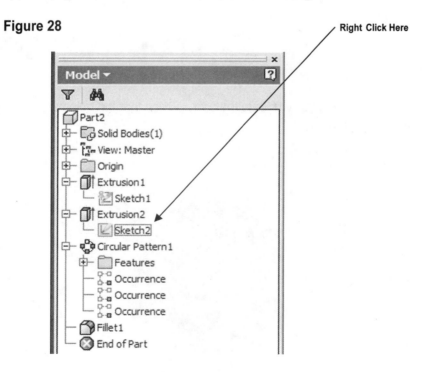

34. A pop up menu will appear. Left click on **Edit Sketch** as shown in Figure 29.

Figure 29

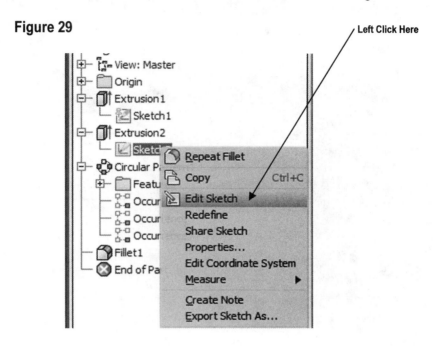

35. The original sketch will appear as shown in Figure 30.

Figure 30

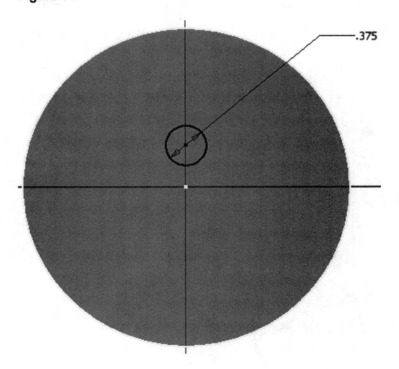

36.   Modify the diameter of the holes by double clicking on the overall dimension. The Edit Dimension dialog box will appear as shown in Figure 31.

Figure 31

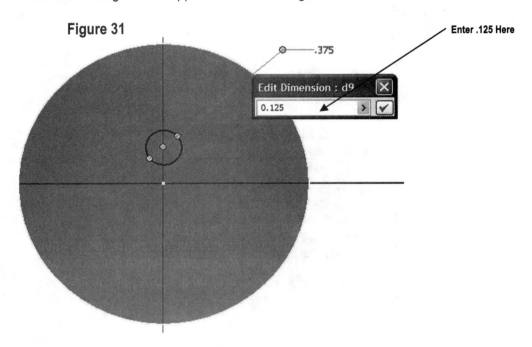

37.   Enter **.125** and press **Enter** on the keyboard as shown in Figure 31.

38.   The diameter of all the holes will be reduced to .125 as shown in Figure 32.

Figure 32

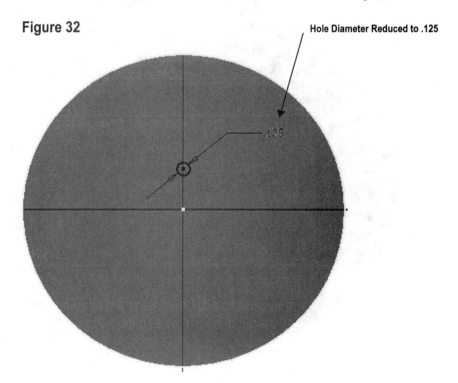

39. Move the cursor to the upper left portion of the screen and left click on the **Manage** tab. Left click on the drop down arrow below Update. A drop down menu will appear. Left click on **Update** as shown in Figure 33.

Figure 33

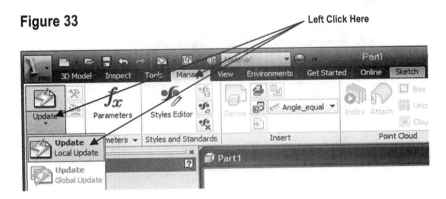

40. Inventor will automatically update the part as shown in Figure 34.

Figure 34

41. Move the cursor over the text "Extrusion2". A red box will appear around the text. After the red box appears, right click once on **Extrusion2** as shown in Figure 35.

Figure 35

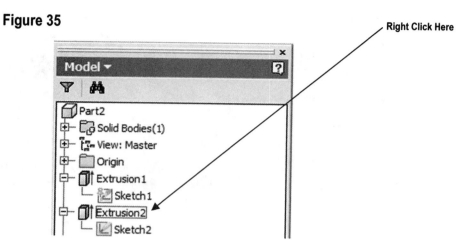

42. A pop up menu will appear. Left click on **Edit Feature** as shown in Figure 36.

Figure 36

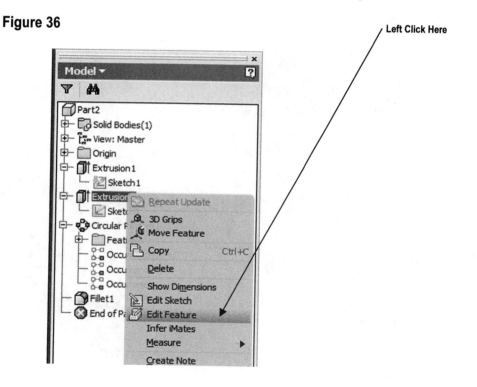

43. The Extrusion dialog box will appear. Enter **.5** for the extrusion distance and left click on **OK** as shown in Figure 37.

Figure 37

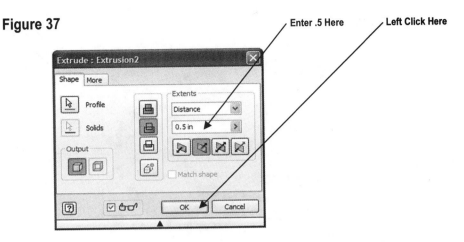

44. Move the cursor to the upper left portion of the screen and left click on the **Manage** tab. Left click on the drop down arrow below Update. A drop down menu will appear. Left click on **Update** as shown in Figure 38.

Figure 38

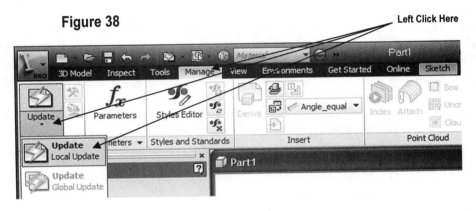

45. Inventor will automatically update the part. Notice that the holes are now thru holes as shown in Figure 39.

Figure 39

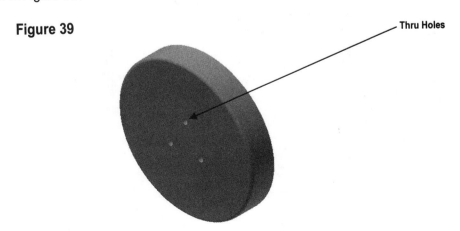

46. Use the Free Orbit/Rotate command to rotate the part as shown in Figure 40.

Figure 40

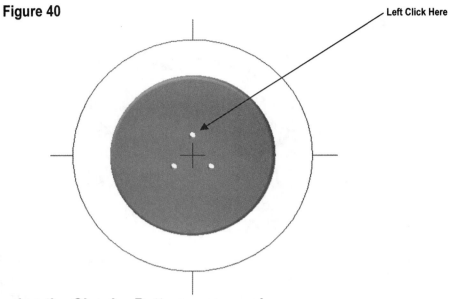

Left Click Here

Edit the part using the Circular Pattern command

47. Move the cursor over the text "Circular Pattern1". A red box will appear around the text. After the red box appears, left click once on **Circular Pattern 1** as shown in Figure 41.

Figure 41

Right Click Here

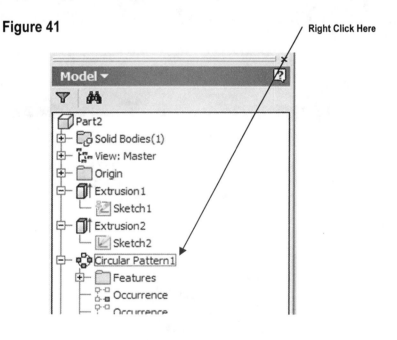

48. Right click on **Circular Pattern1**. A pop up menu will appear. Left click on **Edit Feature** as shown in Figure 42.

Figure 42

Left Click Here

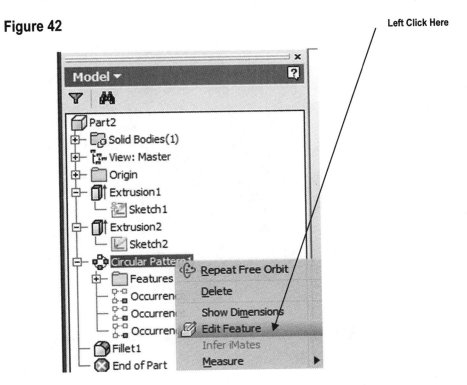

49. The Circular Pattern dialog box will appear. Enter **6** under "Placement" as shown in Figure 43.

Figure 43

Enter 6 Here

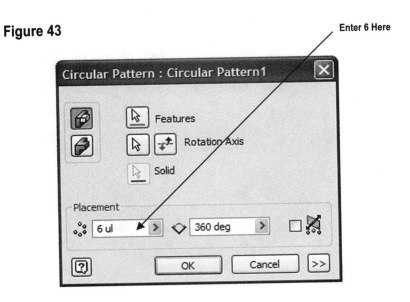

50.   Inventor will provide a preview as shown in Figure 44.

Figure 44

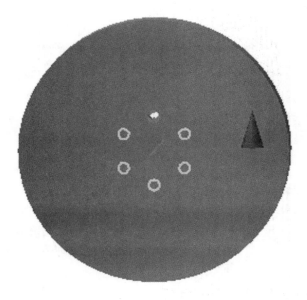

51.   Left click on **OK** in the Circular Pattern dialog box. Your screen should look similar to Figure 45.

Figure 45

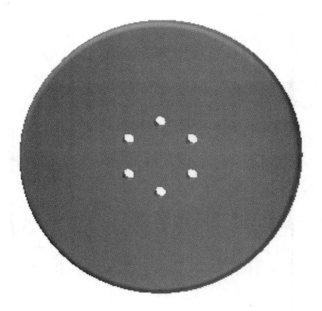

52.   Change the view to **Home/Isometric View** as shown in Figure 46.

Figure 46

Edit the part using the Fillet command

53.   Move the cursor over the text "Fillet1". A red box will appear around the text. After the red box appears, left click once on **Fillet1** as shown in Figure 47.

Figure 47

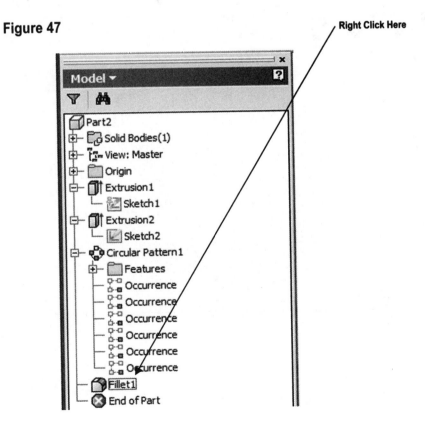

Right Click Here

54. Right click on **Fillet1**. A pop up menu will appear. Left click on **Edit Feature** as shown in Figure 48.

Figure 48

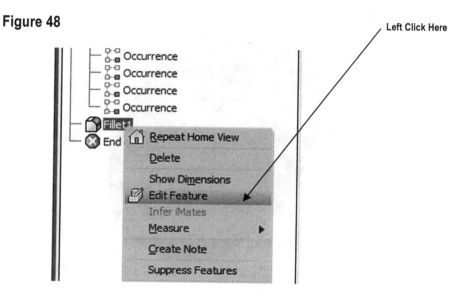

55. The Fillet dialog box will appear. Enter **.250** for the Radius and left click on **OK** as shown in Figure 49.

Figure 49

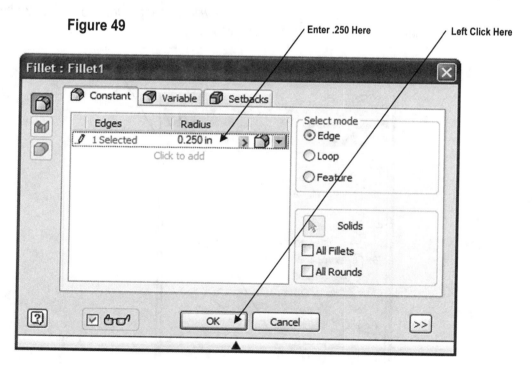

56. Your screen should look similar to Figure 50.

Figure 50

57. Move the cursor over the second listed text "Occurrence". A red box will appear around the text. After the red box appears, left click once on **Occurrence** as shown in Figure 51.

Figure 51

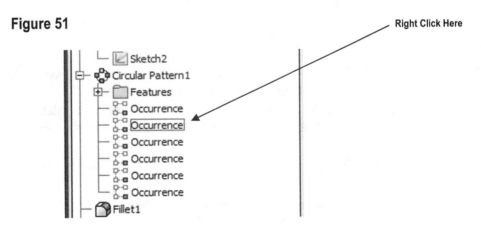

58.   Right click once. A pop up menu will appear. Left click on **Suppress** as shown in Figure 52. Inventor will suppress that particular occurrence while leaving all others active.

Figure 52

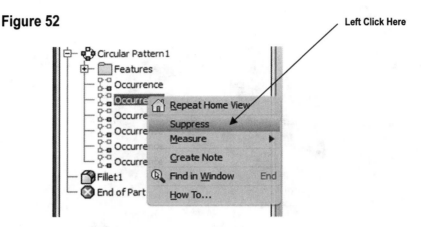

59.   Inventor will draw a line through and gray the text as shown in Figure 53. You will notice that the second hole created using the Circular Pattern command is not visible. Repeat the previous steps to un-suppress the occurrence.

Figure 53

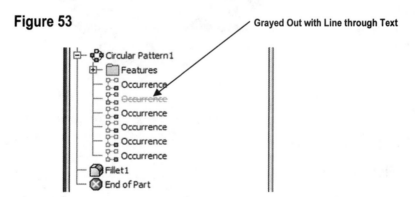

60.   The names of all branches in the part tree can also be edited. Move the cursor to the lower left portion of the screen where the part tree is located. Move the cursor over **Extrusion1** and left click once causing the text to become highlighted. After the text is highlighted, left click one time. The text may be edited as shown in Figure 54.

Figure 54

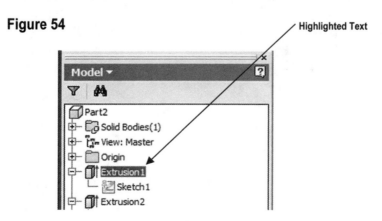

61. Enter the text **Base Extrusion** as shown in Figure 55. Press **Enter** on the keyboard. Text for each individual operation can be edited if desired.

Figure 55

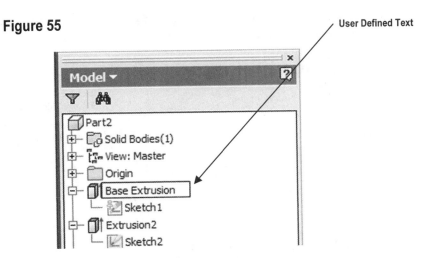

62. Notice that the final design looks significantly different than the original design. The part was redesigned by modifying the existing part as shown in Figure 56.

Figure 56

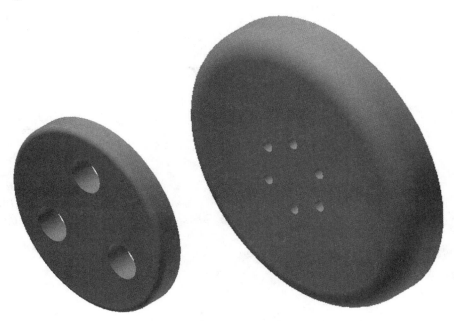

Chapter Problems

Create the following parts as they are shown. Then use the directions to the right and modify the part accordingly. Use the Sketch Panel and Features Panel to modify/edit each part.

Problem 5-1

Use the Extrude command to modify the extrusion thickness to 0.125 inches

Use the Edit Sketch command to modify the sketch to 3.00 inches

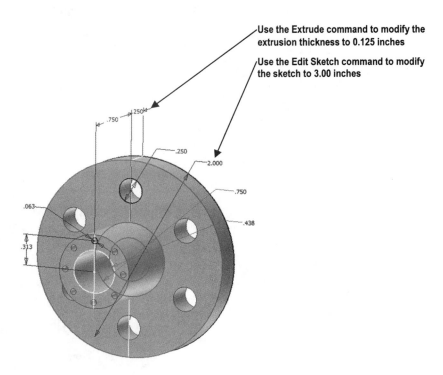

Problem 5-2

Use the Extrude command to modify the Extrusion distance to 2.00 inches. This will cause the hub to protrude out of the back of the part.

Use the Edit Sketch command to modify the sketch to 1.00 inch

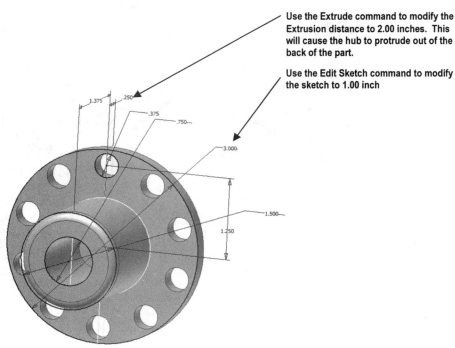

Problem 5-3

Use the Edit Sketch command to increase the hole diameter to .375

Use the Edit Sketch command to increase the hole center distance to 1.25 inches

Use the Circular Pattern command to increase the number of holes to 10

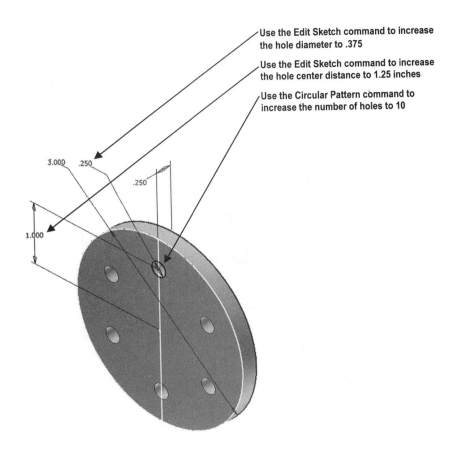

Problem 5-4

Use the Extrude command to increase the distance to 1 inch

Use the Extrude-Cut command to increase the cut distance to 1 inch

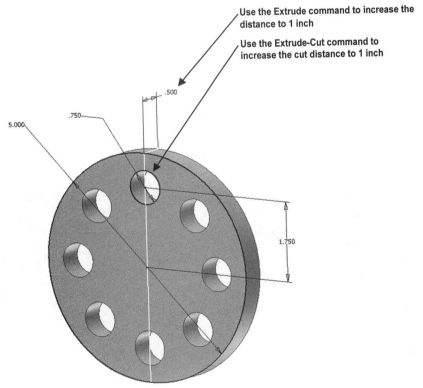

Problem 5-5

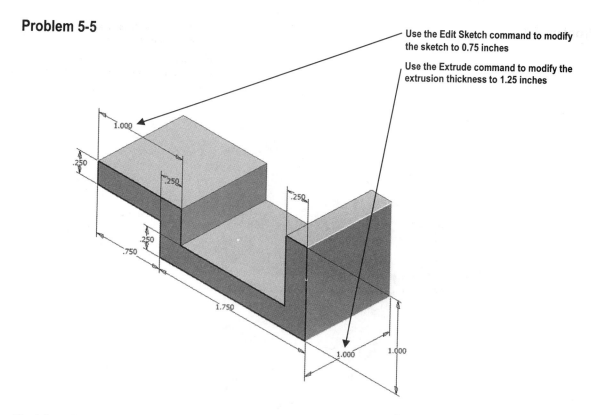

Use the Edit Sketch command to modify the sketch to 0.75 inches

Use the Extrude command to modify the extrusion thickness to 1.25 inches

Problem 5-6

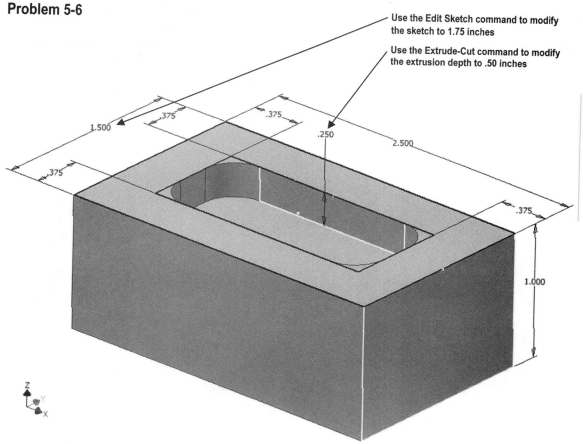

Use the Edit Sketch command to modify the sketch to 1.75 inches

Use the Extrude-Cut command to modify the extrusion depth to .50 inches

Problem 5-7

Use the Edit Sketch command to modify the sketch to 0.875 inches

Use the Extrude-Cut command to modify the Extrusion to a Reverse Direction Cut

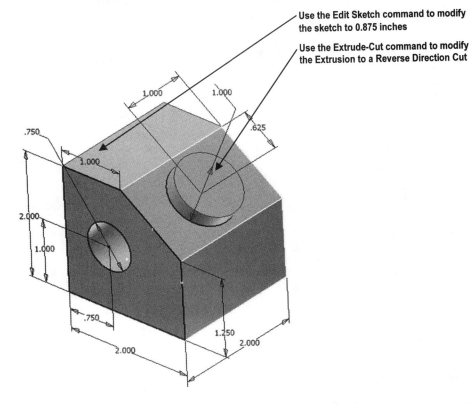

Problem 5-8

Use the Edit Sketch command to modify the sketch to 1.275 inches

Use the Extrude-Cut command to modify the Extruded Cut Depth to .25 inches

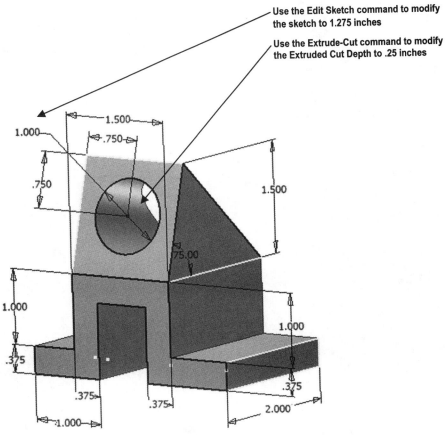

Notes:

CHAPTER 6

Designing Part Models for Assembly

Objectives:

1. Design multiple sketch parts
2. Learn to use the X, Y, and Z Planes
3. Learn to use the Wireframe viewing command
4. Learn to project geometry on to a new sketch
5. Learn to use the Shell command
6. Learn to use Constraints while constructing a Sketch

Chapter 6 includes instruction on how to design the parts shown.

1. Start Autodesk Inventor 2014 by referring to "Chapter 1 Getting Started".

2. After Autodesk Inventor 2014 is running, begin a new sketch.

3. Complete the sketch shown in Figure 1.

    Figure 1

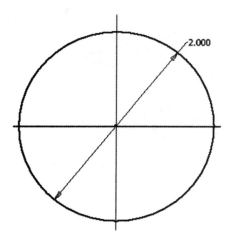

4. Exit out of the Sketch Panel and change the view to Isometric as shown in Figure 2.

    Figure 2

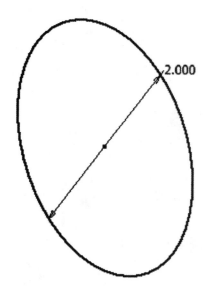

5. Extrude the sketch a distance of 2 inches as shown in Figure 3.

Figure 3

6. Move the cursor to the upper left portion of the screen and left click on the plus sign next to the text "Origin" as shown in Figure 4.

Figure 4

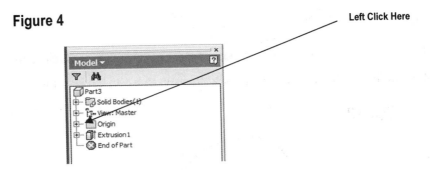

Learn to use the X, Y, and Z Planes

7. The part tree will expand. Move the cursor over the text "YZ Plane" causing a red box to appear around the text as shown in Figure 5.

Figure 5

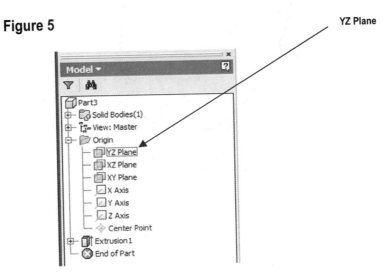

8. The YZ plane will become visible as shown in Figure 6.

Figure 6

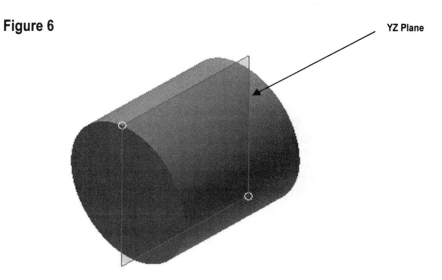

YZ Plane

Learn to use the Wireframe viewing command

9. Right click on the text **YZ Plane**. A pop up menu will appear. Left click on **New Sketch** as shown in Figure 7.

Figure 7

Left Click Here

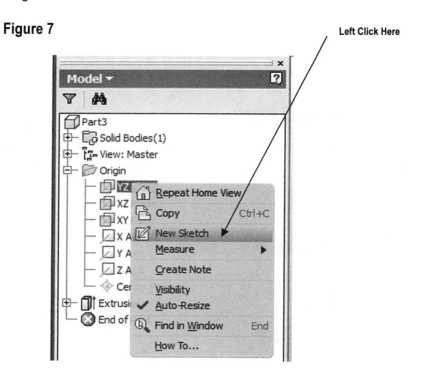

10. Use the Home View command to rotate the part as shown. Your screen should look similar to Figure 8.

Figure 8

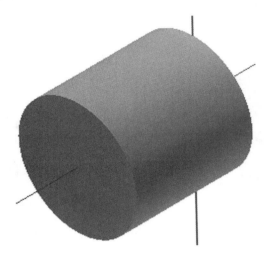

Learn to use the Wireframe viewing command

11. Move the cursor to the upper left portion of the screen and left click on the **View** tab. Left click on the drop down arrow to below "Visual Style" icon. A drop down menu will appear. Left click on **Wireframe** as shown in Figure 9.

Figure 9

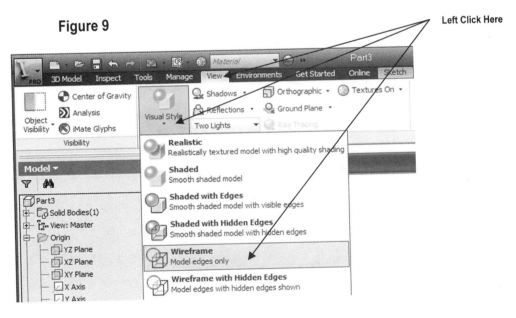

12. Your screen should look similar to Figure 10.

Figure 10

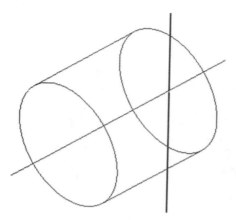

13. Move the cursor to the upper right portion of the screen and left click on the "View Face/ Look At" icon as shown in Figure 11.

Figure 11　　　　　　　　　　　　　　　Left Click Here

14. Move the cursor to the upper left portion of the screen and left click on the text **YZ Plane** in the part tree as shown in Figure 12.

Figure 12　　　　　　　　　　　　　　　Left Click Here

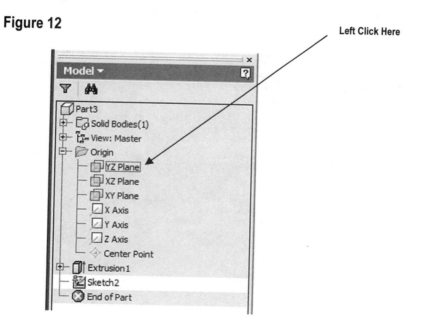

15. Notice the YZ plane becoming visible through the part as shown in Figure 13.

Figure 13

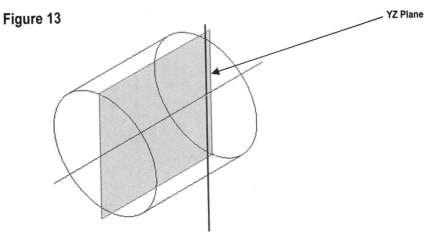

YZ Plane

Learn to project geometry to a new sketch

16. Inventor will rotate the YZ plane to provide a perpendicular view as shown in Figure 14.

Figure 14

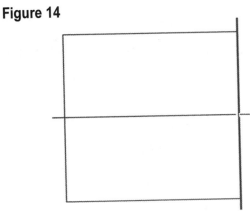

17. Move the cursor to the upper middle portion of the screen and left click on the **Sketch** tab. Left click on **Project Geometry** as shown in Figure 15.

Figure 15

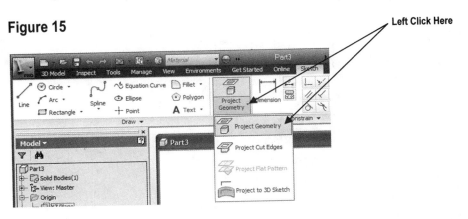

Left Click Here

18. Move the cursor over the top and bottom lines and left click. Inventor will project these lines on to the sketch for reference purposes as shown in Figure 16. They will need to be deleted before exiting the sketch

Figure 16

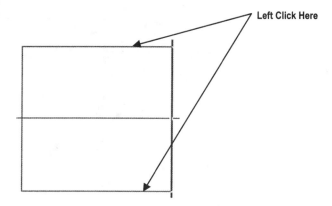

Left Click Here

19. Use the Line command to draw a line from the midpoint of the top line to the midpoint of the bottom line as shown in Figure 17.

Figure 17

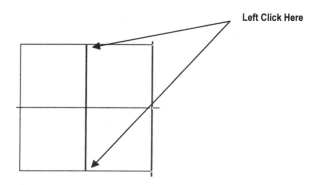

Left Click Here

20. Create a .500 inch circle at the midpoint as shown in Figure 18.

Figure 18

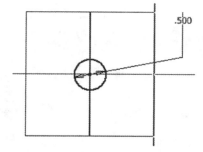

.500

21. Delete the lines that were projected on to the sketch along with the center line that was used to create the circle as shown in Figure 19.

Figure 19

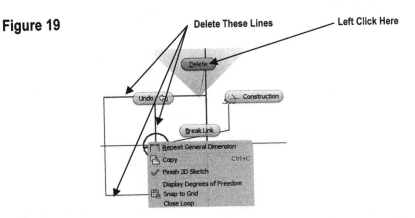

22. Exit out of the Sketch Panel and change the view to Isometric as shown in Figure 20.

Figure 20

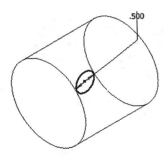

23. Use the Extrude command to create a hole in the part using the newly created circle. Left click on the "Cut" icon. Enter **2.00** for the distance. Left click on the "Bi-directional" icon. Inventor will provide a preview of the extrusion. Left click on **OK** as shown in Figure 21.

Figure 21

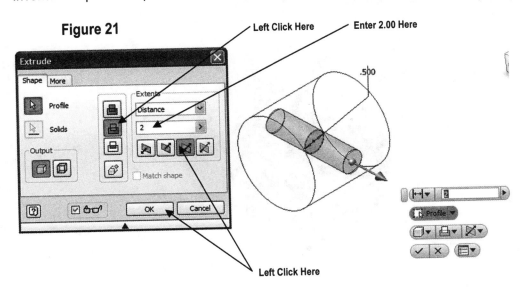

24. Your screen should look similar to Figure 22.

Figure 22

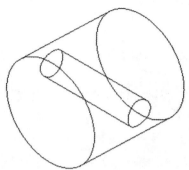

25. Move the cursor to the upper middle portion of the screen and left click on the **View** tab. Left click on the drop down arrow underneath "Visual Style". Left click on **Shaded** as shown in Figure 23.

Figure 23

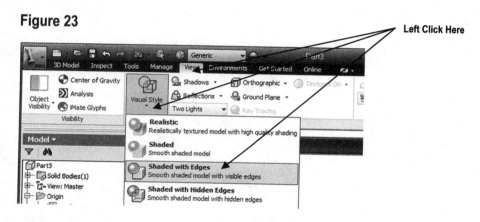

26. Your screen should look similar to Figure 24.

Figure 24

Learn to use the Shell command

27. Move the cursor to the upper left portion of the screen and left click on the Model tab. Left click on the "Shell" icon. The Shell dialog box will appear as shown in Figure 25.

Figure 25

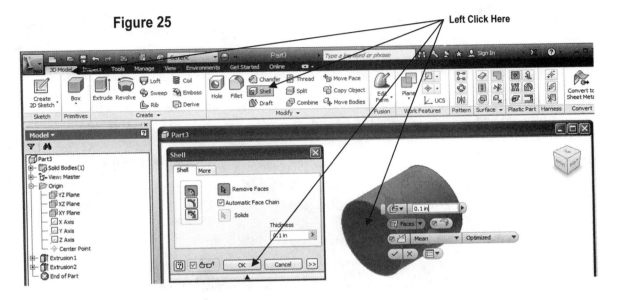

28. Left click on the lower surface of the part and left click on **OK** as shown in Figure 25.

29. Your screen should look similar to Figure 26.

Figure 26

30. Move the cursor to the upper middle portion of the screen and left click on the **View** tab. Left click on the "Face View/Look At" icon as shown in Figure 27.

Figure 27

Left Click Here

31. Move the cursor to the lower surface of the part causing the inside and outside edges to turn red and left click as shown in Figure 28.

Figure 28

Left Click Here

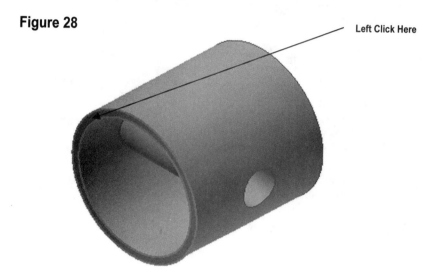

32. After both edges turn red, left click once. Inventor will rotate the part providing a perpendicular view of the inside as shown in Figure 29.

Figure 29

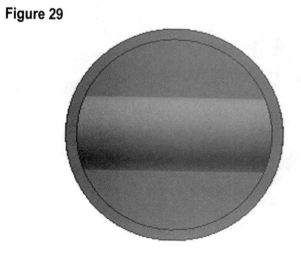

33. Begin a new sketch on the surface shown in Figure 30.

Figure 30

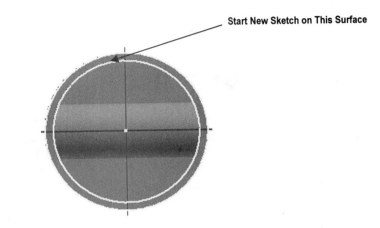

Start New Sketch on This Surface

34. Use the rectangle command to complete the following sketch. You will need to dimension the rectangle from the origin as shown in Figure 31.

Figure 31

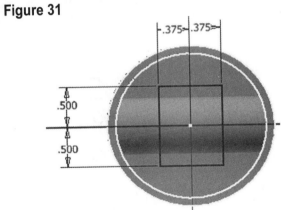

35. Exit the Sketch Panel and change the view to Isometric as shown in Figure 32.

Figure 32

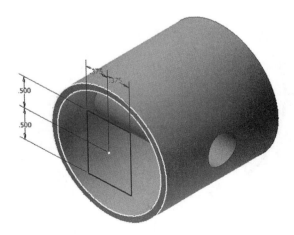

36. Use the Extrude command to "Cut" back into the part a distance of 1.875 as shown in Figure 33.

Figure 33

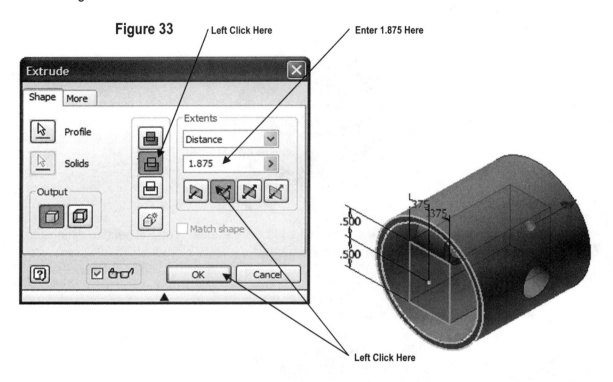

37. Save the part as Piston1.ipt where it can be easily retrieved later.

38. Begin a new drawing as described in Chapter 1.

39. Draw a circle in the center of the grid as shown in Figure 34.

Figure 34

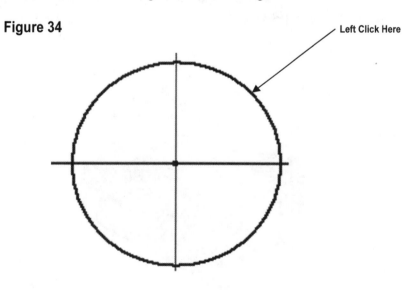

40. Use the **Dimension** command to dimension the circle to **.5** inches as shown in Figure 35.

Figure 35

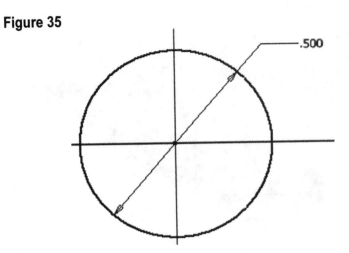

41. Use the **Home View** command to view the sketch in Isometric. Exit the Sketch Panel and Extrude the circle to a length of **1.875** inches as shown in Figure 36.

Figure 36

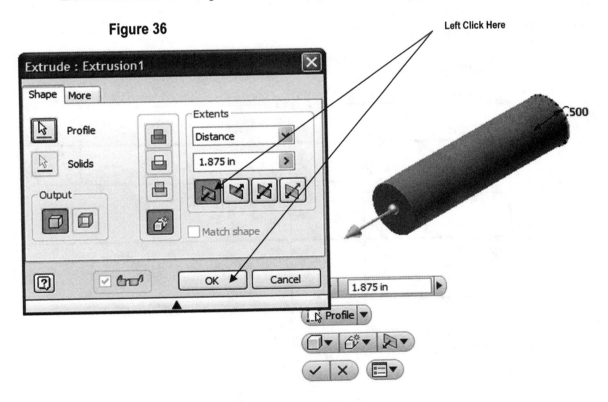

42.   Your screen should look similar to Figure 37.

Figure 37

43.   Save the part as Wristpin1.ipt where it can be easily retrieved later.

44.   Begin a new sketch as described in Chapter 1.

45.   Complete the sketch shown in Figure 38.

Figure 38

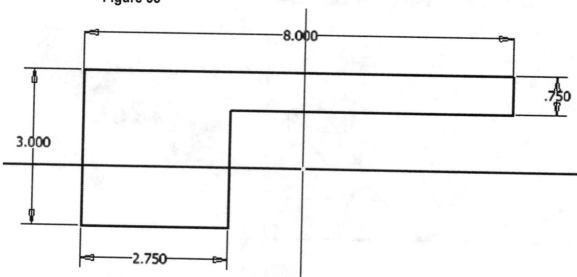

46. Exit the Sketch Panel and change the view to Isometric as shown in Figure 39.

Figure 39

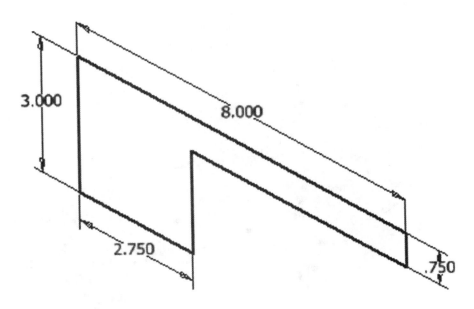

47. Extrude the sketch to a distance of **2.25** inches. Your screen should look similar to what is shown in Figure 40.

Figure 40

48.   Use the Fillet command to create **1.125** inch fillets on the front portion of the part as shown in Figure 41.

Figure 41

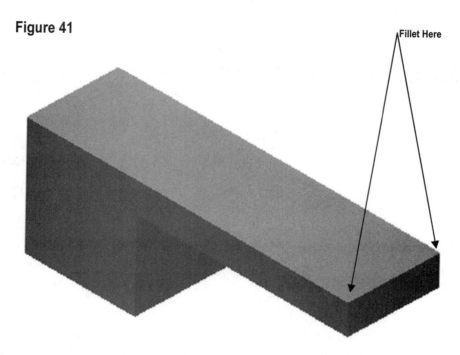

49.   Your screen should look similar to Figure 42.

Figure 42

50. Move the cursor to the upper middle portion of the screen and left click on the **View** tab. Left click on the "Face View/Look At" icon as shown in Figure 43.

Figure 43

51. Move the cursor to the surface shown in Figure 44 causing it to turn red. Left click once.

Figure 44

Move Cursor Here and Left Click

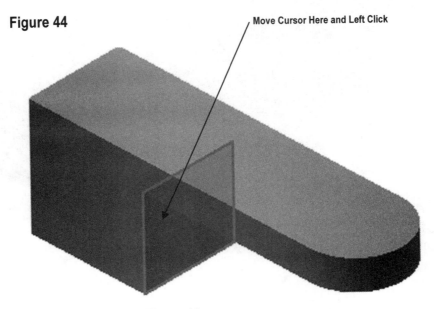

52. Complete the sketch shown in Figure 45.

Figure 45

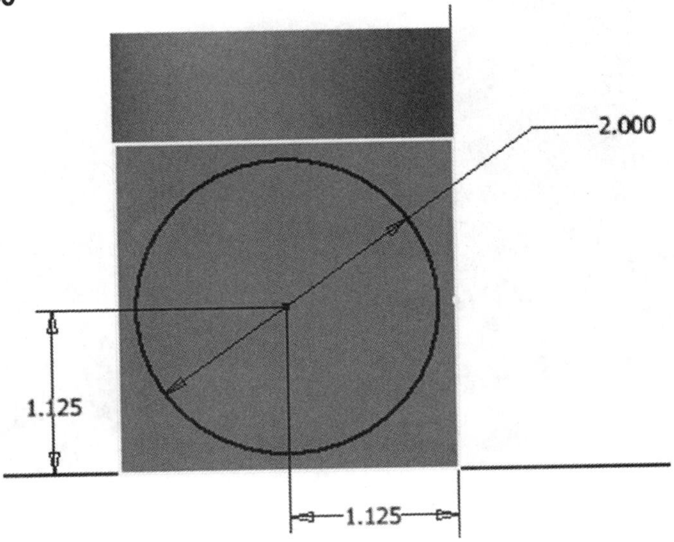

2.000

1.125

1.125

53. Change the view to Isometric as shown in Figure 46.

Figure 46

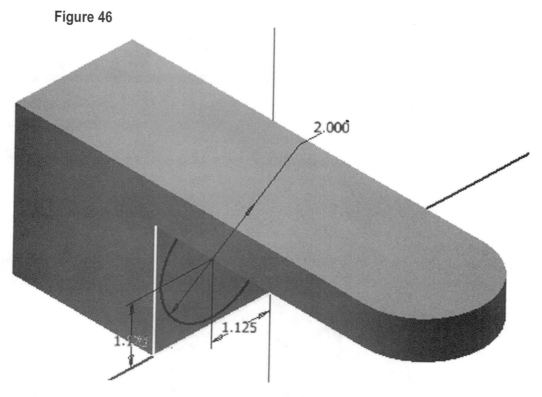

54. Use the Extrude command to extrude or cut out the circle that was just completed creating a thru hole. Your screen should look similar to Figure 47.

Figure 47

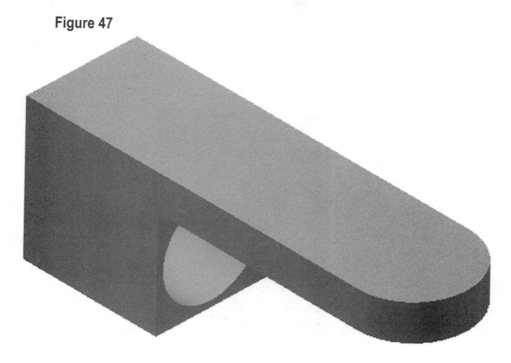

55. Move the cursor to the upper middle portion of the screen and left click on the **View** tab. Left click on the "Face View/Look At" icon as shown in Figure 48.

Figure 48

56. Left click on the surface shown in Figure 49.

Figure 49

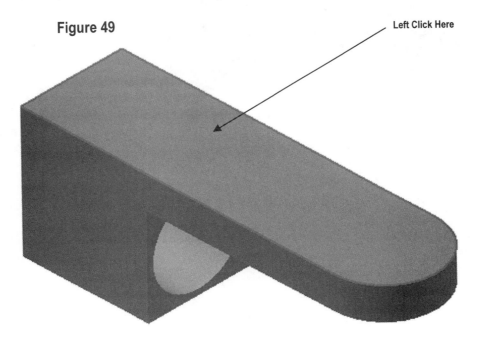

57. Complete the sketch shown in Figure 50.

Figure 50

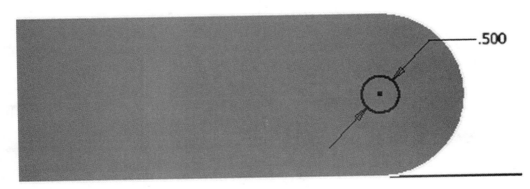

.500

58. Use the Extrude command to extrude or cut out the circle that was just completed. Change the view to Isometric as shown in Figure 51.

Figure 51

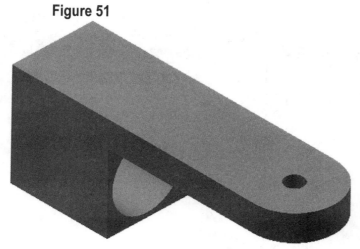

59. Save the part as Pistoncase1.ipt where it can be easily retrieved later.

60. Begin a new drawing as described in Chapter 1.

61. Begin a sketch as shown in Figure 52. Make sure that the circles are located on the endpoint of a line. Also make sure the circles are NOT the same diameter.

Figure 52

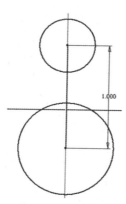

1.000

62. Move the cursor to the upper middle portion of the screen and left click on the "equal" constraints icon as shown in Figure 53.

Figure 53

Left Click Here

63. Left click on each of the circles as shown in Figure 54.

Figure 54

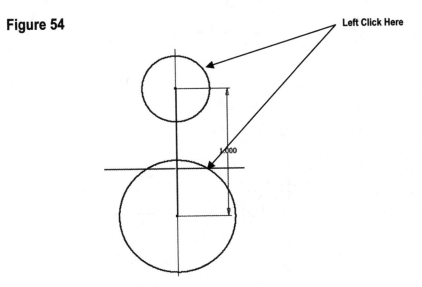

64. Inventor will create two circles of the same size as shown in Figure 55. When one circle is dimensioned, Inventor will automatically update the size of the other circle.

Figure 55

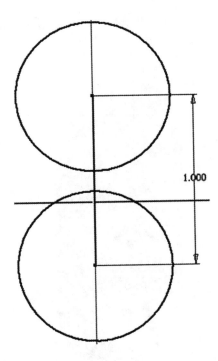

65. Finish completing the sketch shown in Figure 56.

Figure 56

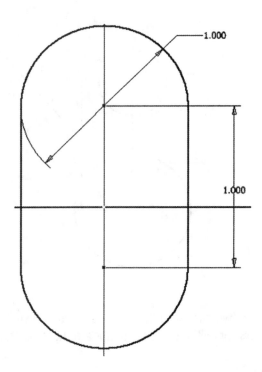

66. Extrude the sketch into a solid with a thickness of **.25** as shown in Figure 57.

Figure 57

67. Complete the following sketch. Use the center of the outside fillet radius as the center of the circle as shown in Figure 58.

Figure 58

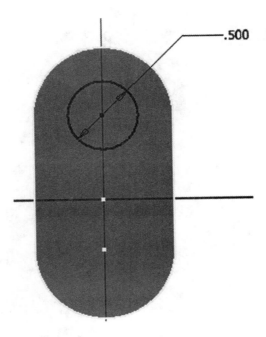

.500

68. Extrude the sketch into a solid with a thickness of **.25** as shown in Figure 59.

Figure 59

69. Rotate the part around to gain access to the opposite side as shown in Figure 60.

Figure 60

70. Complete the following sketch as shown in Figure 61.

Figure 61

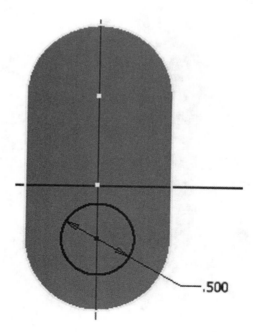

.500

71.   Extrude the sketch into a solid with a thickness of **.25** as shown in Figure 62.

Figure 62

72.   Save the part as Crankshaft1.ipt where it can be easily retrieved later.

73.   Begin a new drawing as described in Chapter 1.

74.   Create a new sketch as shown. Make sure each of the circles are NOT sharing the same center and are NOT in line with each other as shown in Figure 63.

Figure 63

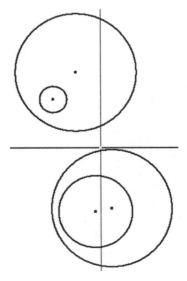

75. Move the cursor to the upper middle portion of the screen and left click on the Concentric constraint icon as shown in Figure 64.

Figure 64

Left Click Here

76. Holding the Shift key down, left click on each circle as shown in Figure 65.

Figure 65

Left Click Here

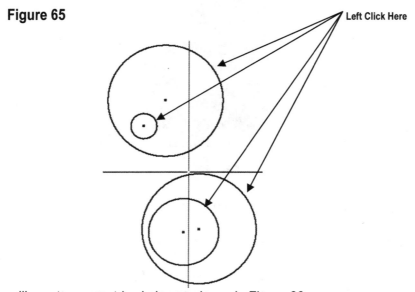

77. Inventor will create concentric circles as shown in Figure 66.

Figure 66

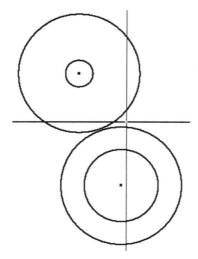

78. Move the cursor to the upper middle portion of the screen and left click on the Vertical constraint icon as shown in Figure 67.

Figure 67

Left Click Here

79. Left click on the centers of the circles as shown in Figure 68.

Figure 68

Left Click Here

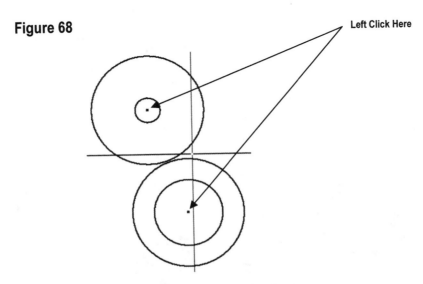

80. Inventor will create a vertical constraint between the centers of the circles as shown in Figure 69.

Figure 69

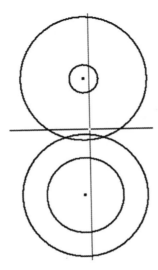

81. Use the Free Orbit/Rotate command to rotate the sketch around on to its side as shown in Figure 70.

Figure 70

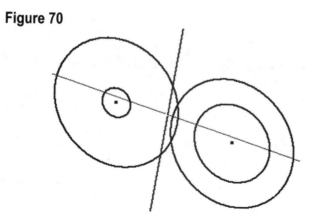

82. Using the geometry created above, complete the sketch shown. Extrude the sketch to a thickness of **.25** inches as shown in Figure 71.

Figure 71

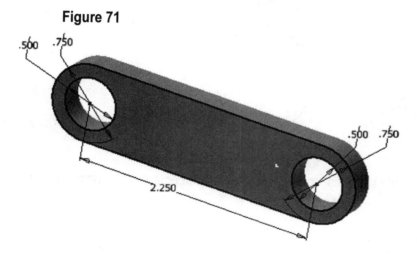

83. Save the part as Conrod1.ipt where it can be easily retrieved later.

84. All of these parts will be used in the next chapter.

CHAPTER 7

Introduction to Assembly View Procedures

Objectives:

1. Learn to import existing solid models into the Assembly Panel
2. Learn to constrain all parts in the Assembly Panel
3. Learn to edit/modify parts while in the Assembly Panel
4. Learn to assign colors to different parts in the Assembly Panel
5. Learn to animate/simulate motion
6. Learn to create an .avi or .wmv file while in the Assembly Panel

Chapter 7 includes instruction on how to construct the assembly shown.

1. Start Inventor 2014 by referring to "Chapter 1 Getting Started".

2. After Autodesk Inventor 2014 is running, begin an Assembly Drawing. First, move the cursor to the upper left corner of the screen and left click on **New**. The Create New File dialog box will appear. Left click on the **English** folder. Left click on **Standard (in).iam** as shown in Figure 1.

Figure 1

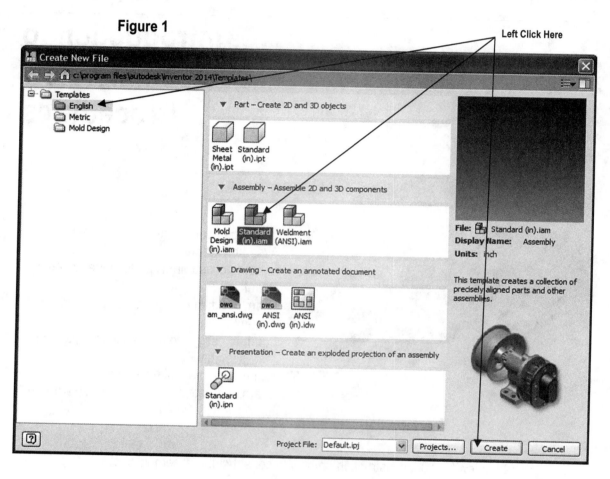

3. Left click on **Create**.

Learn to import existing solid models into the Assemble Panel

4.   The Assemble Panel will open. Your screen should look similar to Figure 2.

Figure 2

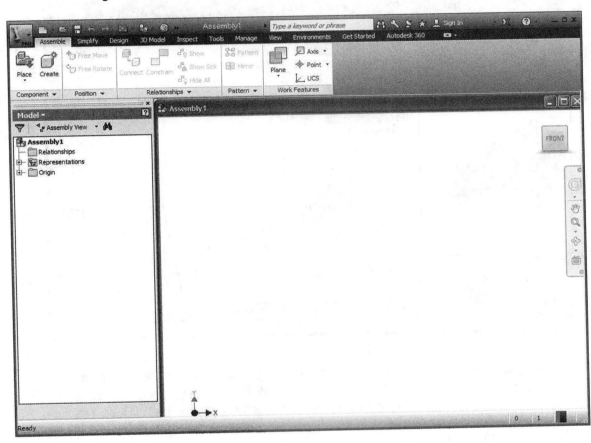

5.   Move the cursor to the upper left portion of the screen and left click on **Place** as shown in Figure 3.

Figure 3

6. The Place Component dialog box will appear. Locate the PistonCase1.ipt file and left click on **Open** as shown in Figure 4.

Figure 4

Left Click Here

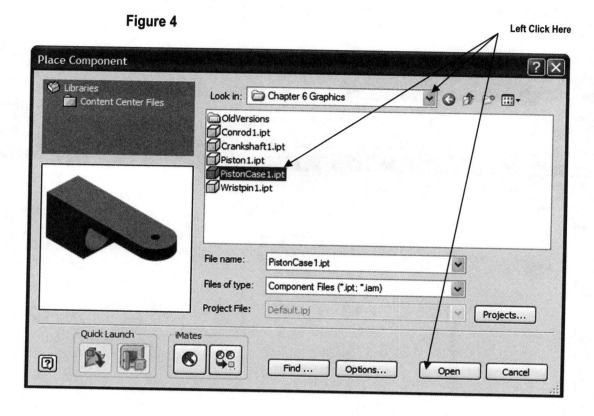

7. Inventor will place one piston case in the drawing space while another piston case will be attached to the cursor as shown in Figure 5. A dialog box may appear indicating that "The location of the selected file drawing in not in the active project". Left click on **Yes**.

Figure 5

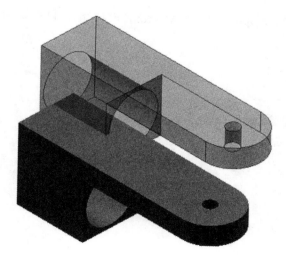

8. Do **NOT** left click. Left clicking would cause Inventor to place two piston cases in the Assembly area. Press the **Esc** key on the keyboard. Your screen should look similar to Figure 6.

Figure 6

9. Move the cursor to the upper left portion of the screen and left click on **Place** as shown in Figure 7.

Figure 7

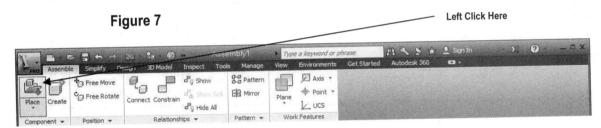

10. The Place Component dialog box will appear. Locate the Piston1.ipt file and left click on **Open** as shown in Figure 8.

Figure 8

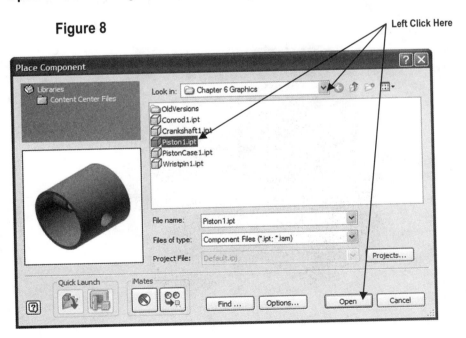

11. The piston will be attached to the cursor. Place the piston anywhere near the piston case and left click once. Another piston will be attached to the cursor in case another will be used. In this drawing there is no need to import the same part multiple times. Press the **Esc** button on the keyboard once. Your screen should look similar to Figure 9.

Figure 9

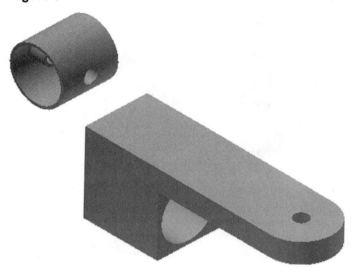

12. Continue to "Place" the remaining parts into the Assembly area as shown in Figure 10.

Figure 10

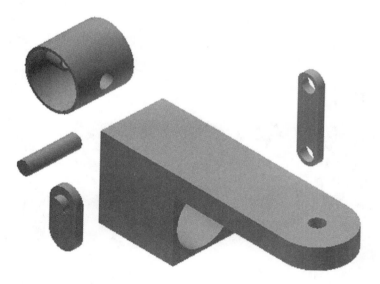

13. Move the cursor to the lower left portion of the screen in the part tree. Notice the picture of a push pin that appears next to the piston case text in the branch of the part tree. Move the cursor over the words **Pistoncase:1** and left click once. The text will turn blue. Right click once. A pop up menu will appear. Left click on **Grounded** as shown in Figure 11. Inventor will "Unground" the case allowing it to be moved using the rotate component command as shown in Figure 12. **The first part placed into any assembly is automatically grounded**.

Figure 11

Push Pin   Left Click then Right Click Here   Left Click Here

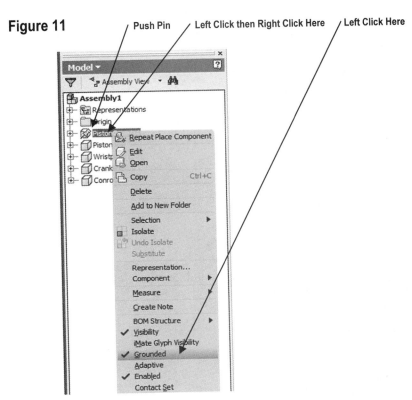

14. Move the cursor to the upper middle portion of the screen and left click on the **Rotate Component** icon as shown in Figure 12.

Figure 12

Left Click Here

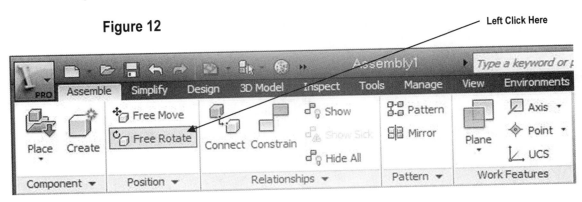

15. Move the cursor to the piston case and left click once. A white circle will appear around the piston case. Rotate the piston case upward as shown in Figure 13.

Figure 13

White Circle

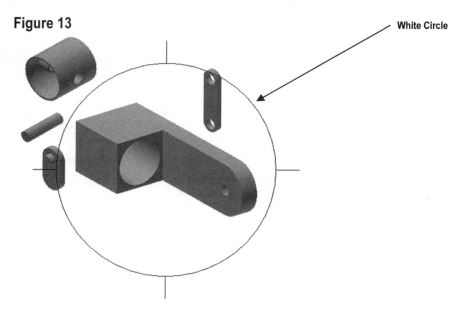

16. Your screen should look similar to Figure 13.

17. After the piston case is rotated as shown in Figure 14, right click once. A pop up menu will appear. Left click on **Done** as shown in Figure 14.

Figure 14

Left Click Here

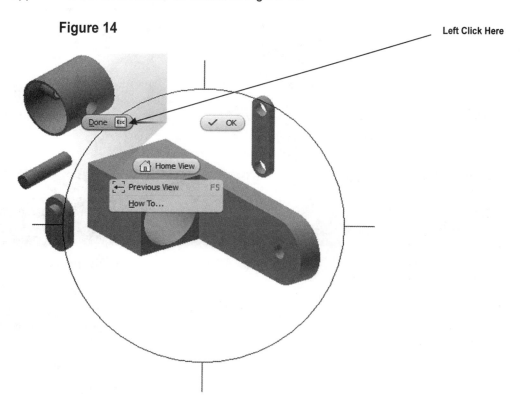

18.   Move the cursor to the lower left portion of the screen in the part tree. Notice the picture of the push pin that appeared next to the piston case text is gone. This means the piston case is NOT grounded. Move the cursor over the text **Pistoncase1:1** and left click once. The text will turn blue. Right click once. A pop up menu will appear. Left click on **Grounded** as shown in Figure 15. Inventor will "ground" the case preventing it from being moved while the rest of the assembly is constructed. **Caution: Only ground the Piston Case. If any other parts inadvertently become grounded it will be impossible to assemble parts together.**

Figure 15

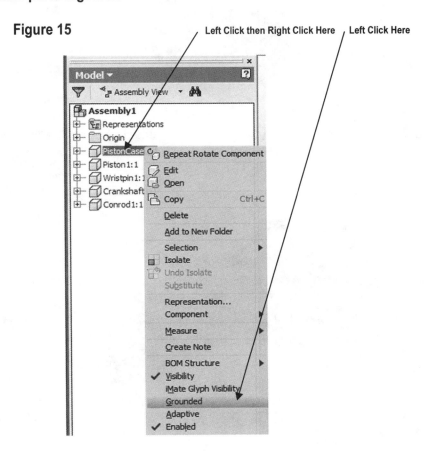

Learn to constrain all parts in the Assemble Panel

19.   Move the cursor to the upper middle portion of the screen and left click on **Constrain**. The Place Constraint dialog box will appear as shown in Figure 16.

Figure 16

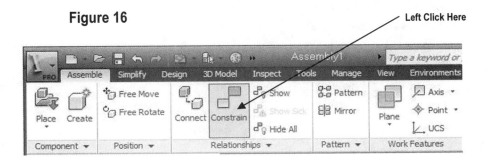

20. Move the cursor over the piston until a red center line appears as shown in Figure 17. Left click once.

Figure 17

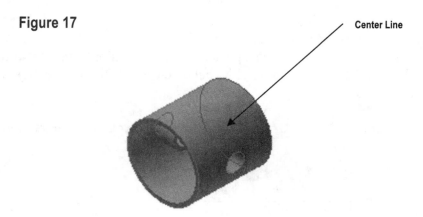

Center Line

21. Move the cursor over the piston case until a red center line appears as shown in Figure 18. Left click once.

Figure 18

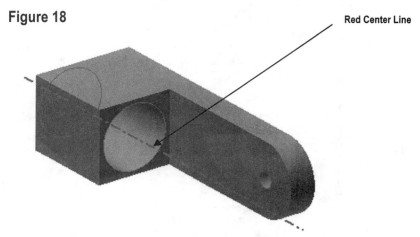

Red Center Line

22. Inventor will align the centers of the piston and the piston case. Your screen should look similar to Figure 19.

Figure 19

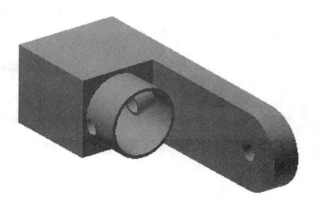

23. If Inventor installed the piston upside down, click on the "Undo" icon. Use the Rotate Component command to rotate the piston so that Inventor has to rotate it less than 180 degrees to install it.

24. Left click on **OK** as shown in Figure 20.

Figure 20

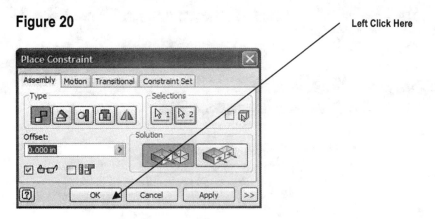

Left Click Here

25. Your screen should look similar to Figure 21.

Figure 21

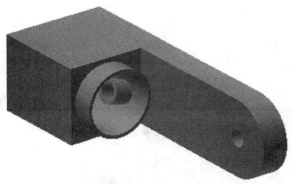

26. Move the cursor to the lower left portion of the piston. Left click (holding the left mouse button down) and slide the piston down out below the bore as shown in Figure 22.

Figure 22

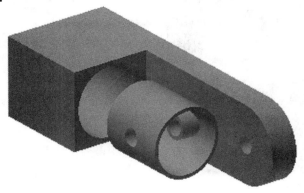

27. Move the cursor to the upper middle portion of the screen and left click on **Constrain**. The Place Constraint dialog box will appear as shown in Figure 23.

Figure 23

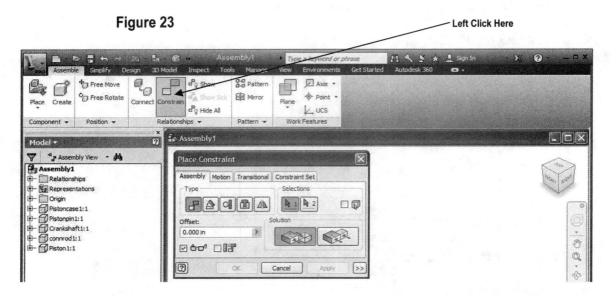

28. Move the cursor to the wristpin hole on the piston. A red center line will appear. Left click once as shown in Figure 24.

Figure 24

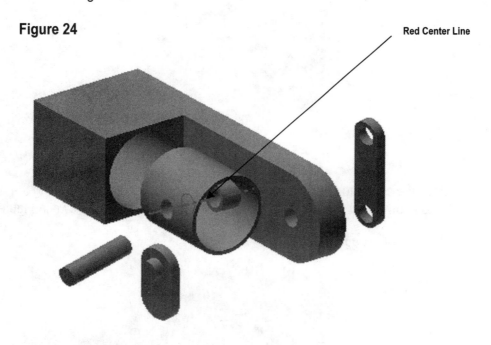

29. Move the cursor to the upper portion of the connecting rod. A red center line will appear. Left click once as shown in Figure 25. You may have to zoom in to accomplish this.

Figure 25

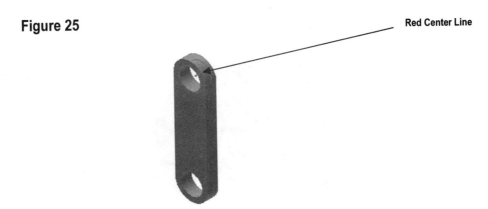

30. Left click on **OK** as shown in Figure 26.

Figure 26

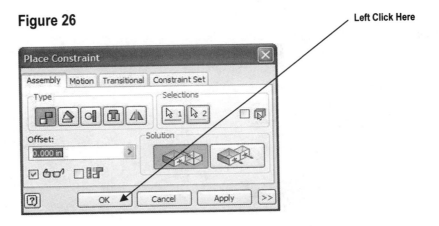

31. Your screen should look similar to Figure 27.

Figure 27

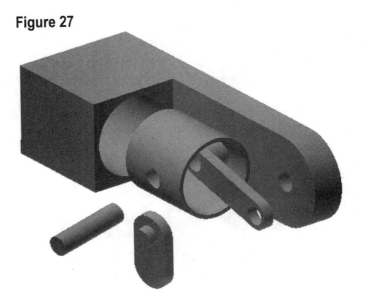

32. Use the Free Orbit/Rotate command to rotate the entire assembly to gain access to the underside of the piston as shown in Figure 28.

Figure 28

33. Move the cursor to the upper middle portion of the screen and left click on **Constrain**. The Place Constraint dialog box will appear as shown in Figure 29.

Figure 29   Left Click Here

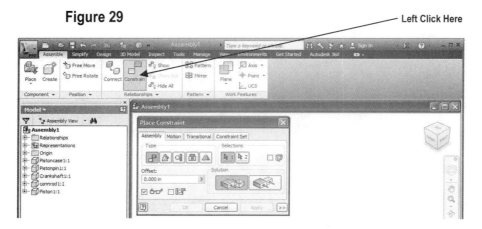

34. Move the cursor to the left side of the connecting rod causing a red arrow to appear. Left click as shown in Figure 30. You may have to zoom in so that Inventor will find the proper surface.

Figure 30   Left Click Here

35. Use the Rotate command to turn the piston in order to gain access to the surface opposite the previously selected surface. Hit the **ESC** key once or right click and select **Done** to get out of the Rotate command. Left click on the surface opposite the previously selected surface as shown in Figure 31.

Figure 31

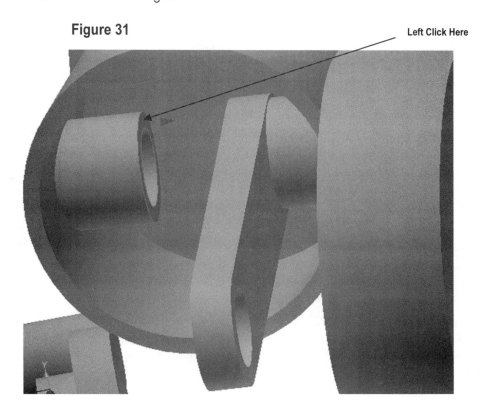

36. Enter **.250** for the offset as shown in Figure 32.

Figure 32

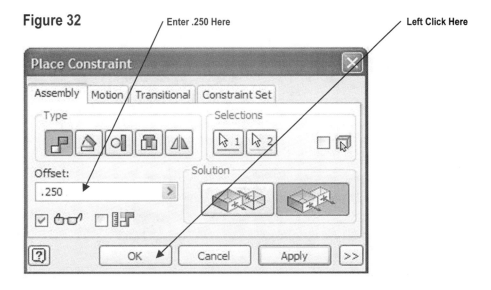

37.   Left click on **OK**.

38.   The connecting rod should be centered in the piston. Your screen should look similar to Figure 33.

Figure 33

39.   Right click anywhere around the drawing. A pop up menu will appear. Left click on **Home View** as shown in Figure 34.

Figure 34

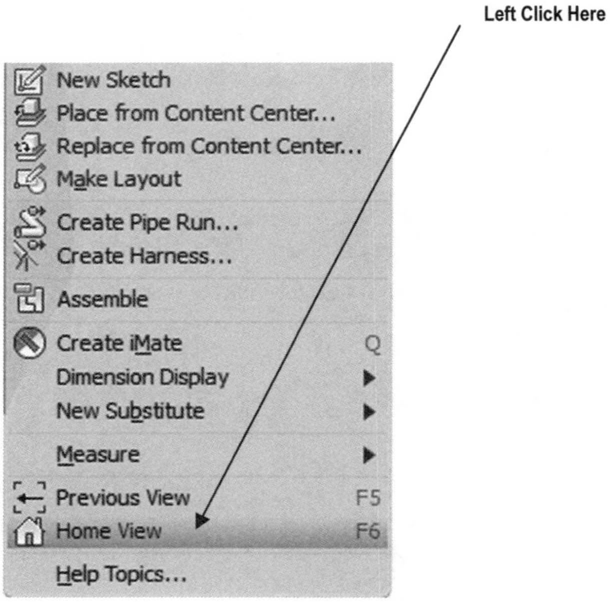

Left Click Here

New Sketch
Place from Content Center...
Replace from Content Center...
Make Layout

Create Pipe Run...
Create Harness...

Assemble

Create iMate   Q
Dimension Display   ▶
New Substitute   ▶

Measure   ▶

Previous View   F5
Home View   F6

Help Topics...

40. Inventor will provide an isometric view of the assembly as shown in Figure 35.

Figure 35

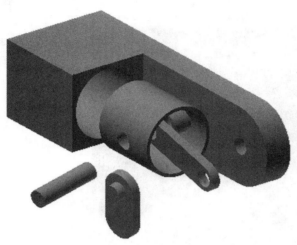

41. Move the cursor to the upper middle portion of the screen and left click on **Constrain**. The Place Constraint dialog box will appear as shown in Figure 36.

Figure 36   Left Click Here

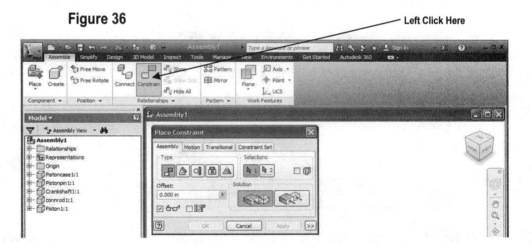

42. Move the cursor to the wrist pin causing a red center line to appear. After a red center line appears, left click once as shown in Figure 37.

Figure 37   Left Click Here

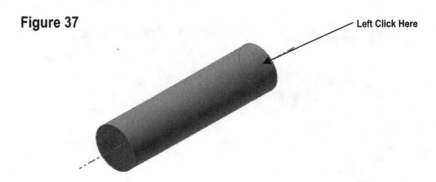

43. Move the cursor to the piston causing a red center line to appear. After a red center line appears, left click once as shown in Figure 38.

Figure 38

Left Click Here

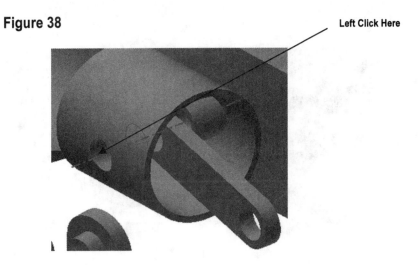

44. Left click on **OK** as shown in Figure 39.

Figure 39

Left Click Here

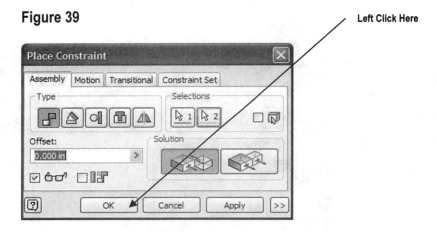

45. Move the cursor to the upper middle portion of the screen and left click on **Constraint**. The Place Constraint dialog box will appear as shown in Figure 40.

Figure 40

Left Click Here

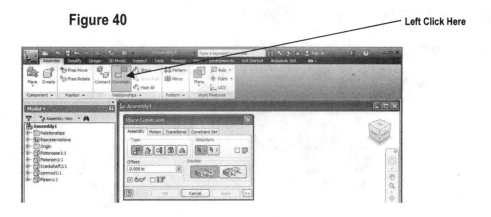

46. Left click on the "Flush" icon as shown in Figure 41.

Figure 41

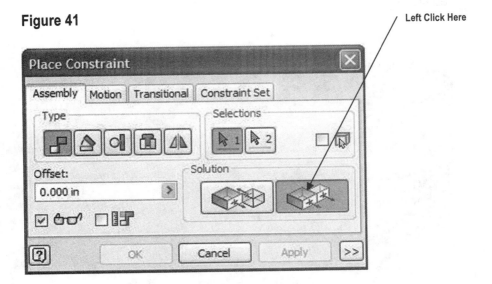

47. Move the cursor to the side of the wrist pin causing a red arrow to appear. After a red arrow appears, left click once as shown in Figure 42.

Figure 42

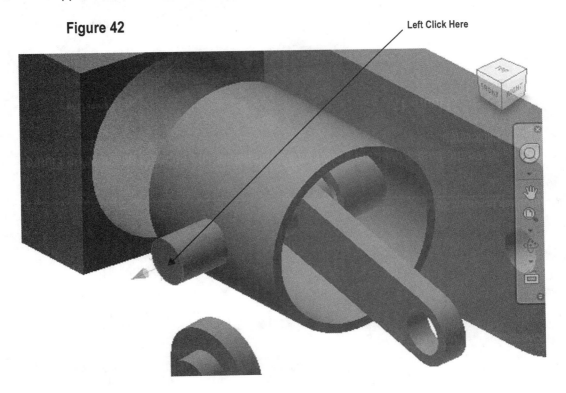

48. Move the cursor to the side of the connecting rod causing a red arrow to appear. After a red arrow appears, left click once as shown in Figure 43.

Figure 43

Left Click Here

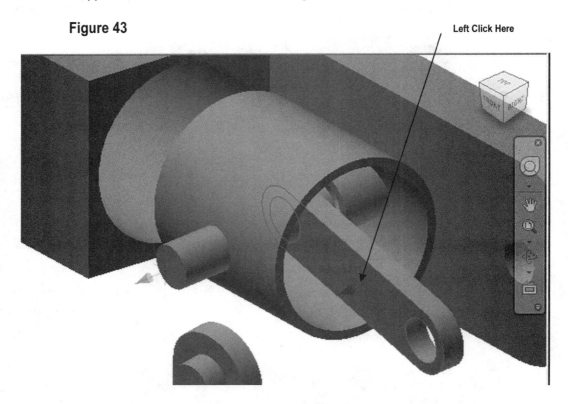

49. Enter **-.7825** under Offset. Left click on **OK** as shown in Figure 44.

Figure 44

Enter -.7825 Here   Left Click Here

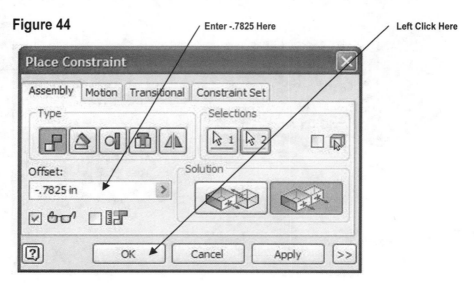

50. Your screen should look similar to Figure 45.

Figure 45

51. Move the cursor to the upper middle portion of the screen and left click on **Constraint**. The Place Constraint dialog box will appear as shown in Figure 46.

Figure 46

Left Click Here

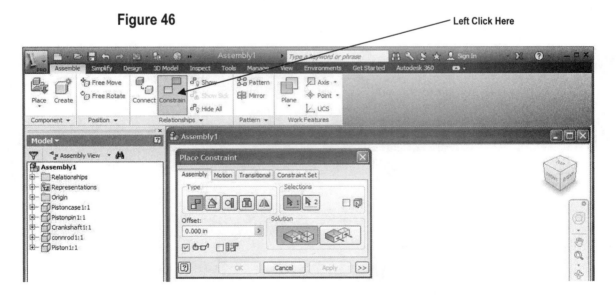

52. Move the cursor to the crankshaft pin causing a red center line to appear. After a red center line appears, left click once as shown in Figure 47. The crankshaft pin will be secured to the connecting rod.

Figure 47

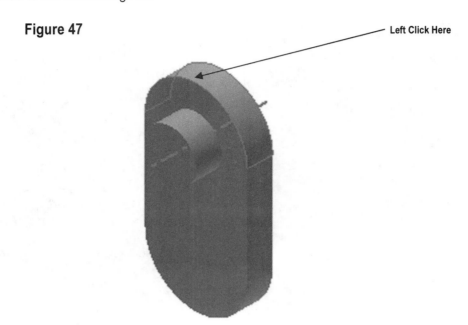

Left Click Here

53. Move the cursor to the connecting rod end causing the red center line to appear. The connecting rod will be secured to the crankshaft. After the red center line appears, left click once as shown in Figure 48.

Figure 48

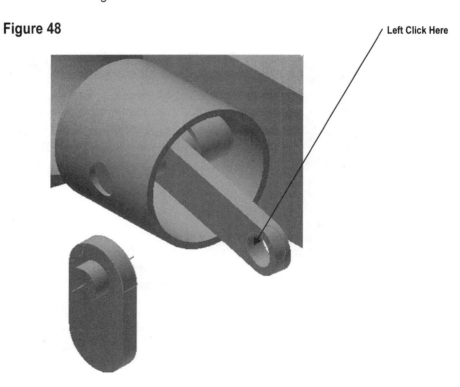

Left Click Here

54. Inventor will place the connecting rod and crankshaft together as shown in Figure 49.

Figure 49

55. Left click on **OK** as shown in Figure 50.

Figure 50

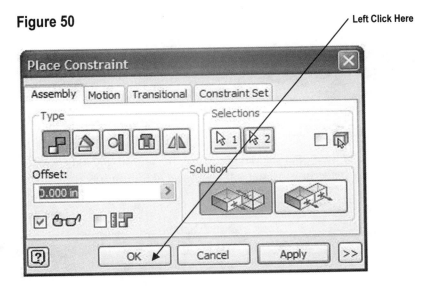

Left Click Here

56. Move the cursor over the piston. Left click (holding the left mouse button down) and drag the piston upward toward the bottom of the bore as shown in Figure 51.

Figure 51

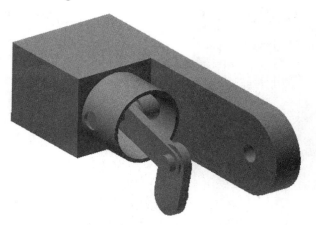

57. Use the Rotate command and roll the assembly around to gain access to the opposite side as shown in Figure 52.

Figure 52

58. Move the cursor to the upper middle portion of the screen and left click on **Constraint**. The Place Constraint dialog box will appear as shown in Figure 53.

Figure 53

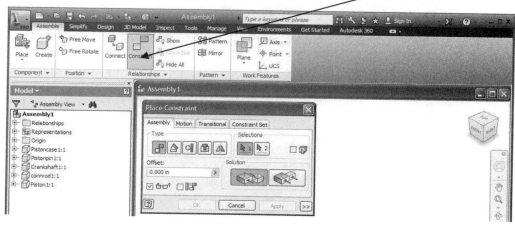

59. Move the cursor to the crankshaft pin, which will be secured in the piston case causing a red center line to appear. After the red center line appears, left click once as shown in Figure 54.

Figure 54

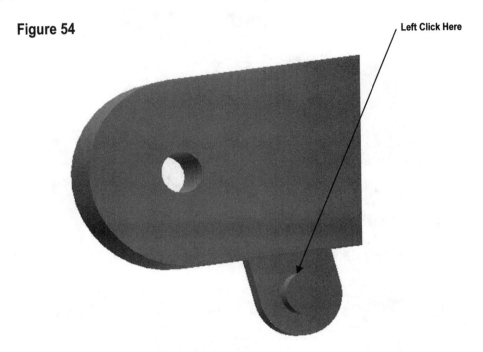

60. Move the cursor to the piston case hole that will secure the crankshaft causing a red center line appear. After the red center line appears, left click once as shown in Figure 55.

Figure 55

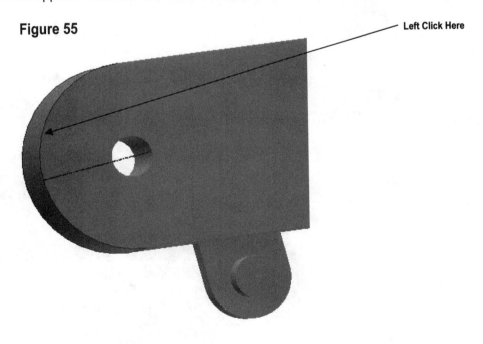

61.   Inventor will place the crankshaft pin into the piston case as shown in Figure 56.

Figure 56

62.   Left click on **OK** as shown in Figure 57.

Figure 57

Left Click Here

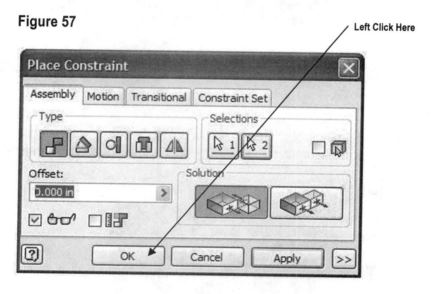

63.   Your screen should look similar to Figure 58.

Figure 58

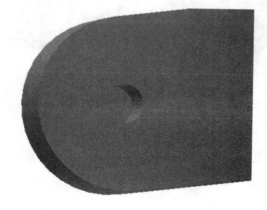

64. Right click anywhere around the drawing. A pop up menu will appear. Left click on **Home View** as shown in Figure 59.

Figure 59

Left Click Here

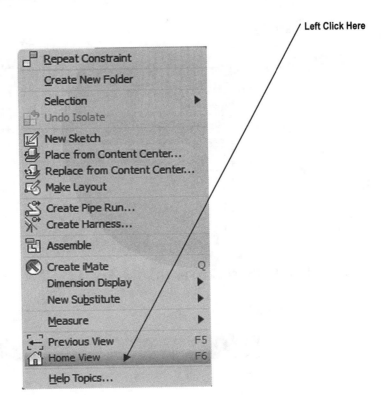

Repeat Constraint

Create New Folder

Selection ▶

Undo Isolate

New Sketch
Place from Content Center...
Replace from Content Center...
Make Layout

Create Pipe Run...
Create Harness...

Assemble

Create iMate   Q
Dimension Display ▶
New Substitute ▶

Measure ▶

Previous View   F5
Home View   F6

Help Topics...

65. Your screen should look similar to Figure 60. If the crankshaft is not visible, it is embedded into the pistoncase. Simply left click on in (holding the left mouse button down) and move the crank out of the pistoncase.

Figure 60

Crankshaft Visible

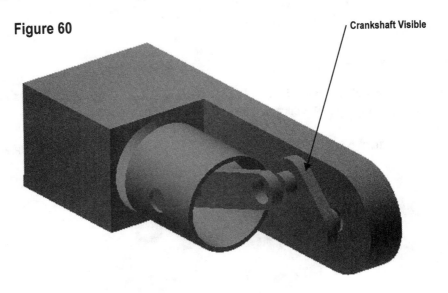

66.   Move the cursor to the upper middle portion of the screen and left click on **Constraint**. The Place Constraint dialog box will appear as shown in Figure 61.

Figure 61

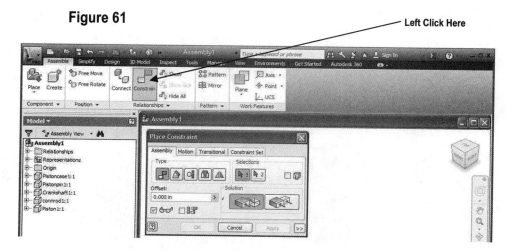

67.   Left click on the "Flush" icon as shown in Figure 62.

Figure 62

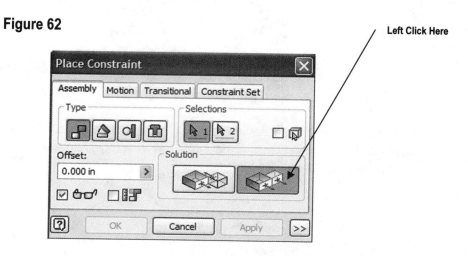

68.   Move the cursor to the left side of the connecting rod causing a red arrow to appear. After a red arrow appears, left click once as shown in Figure 63.

Figure 63

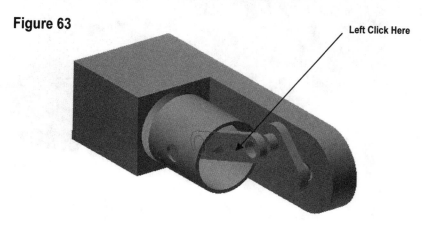

69. Move the cursor to the crankshaft connecting rod pin causing the red arrow to appear. After a red arrow appears, left click once as shown in Figure 64.

Figure 64

Left Click Here

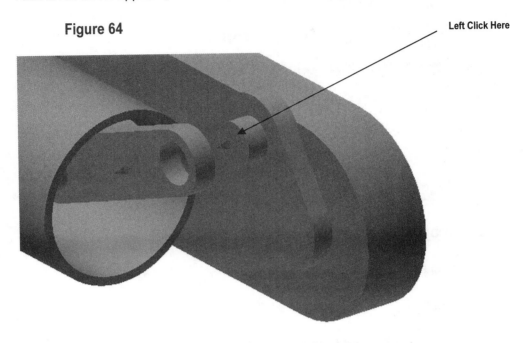

70. Inventor will place the connecting rod flush with the crankshaft connecting rod pin as shown in Figure 65.

Figure 65

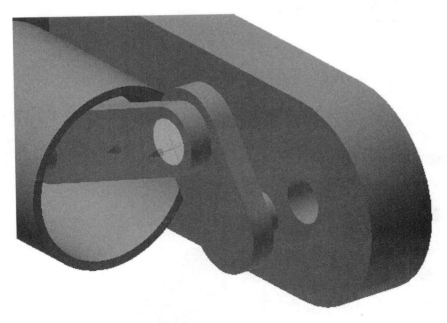

71. Left click on **OK** as shown in Figure 66.

Figure 66

Left Click Here

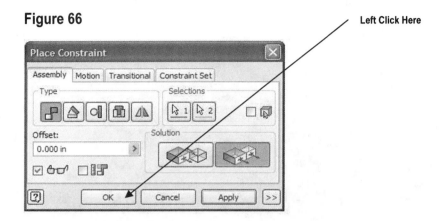

Learn to edit/modify parts while in the Assemble Panel

72. Your screen should look similar to Figure 67.

Figure 67

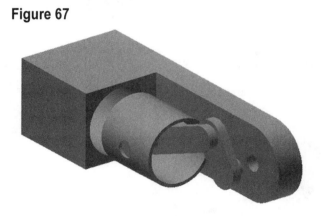

73. The length of the connecting rod must be modified. Move the cursor over the connecting rod causing the edges to turn red as shown in Figure 68.

Figure 68

Move Cursor Here

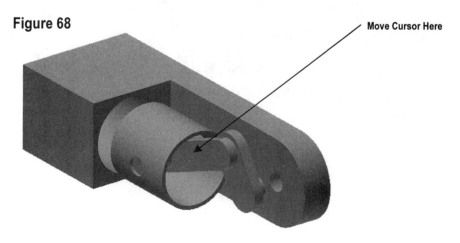

74. Double click (left click) on the connecting rod. All other parts will become grayed as shown in Figure 69.

Figure 69

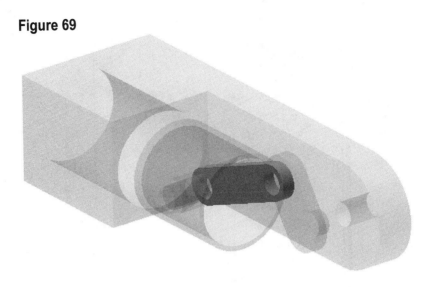

75. Notice that the part tree at the lower left of the screen has changed. All of the branches related to all other parts are grayed (inactive). The branches that illustrate the connecting rod are white (active) as shown in Figure 70.

Figure 70

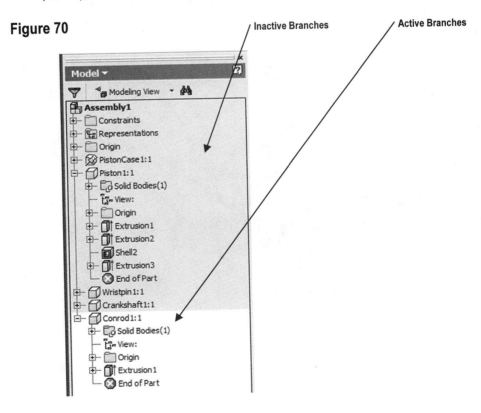

76. Left click on the "Plus" sign next to the text "Extrusion1" as shown in Figure 71.

Figure 71

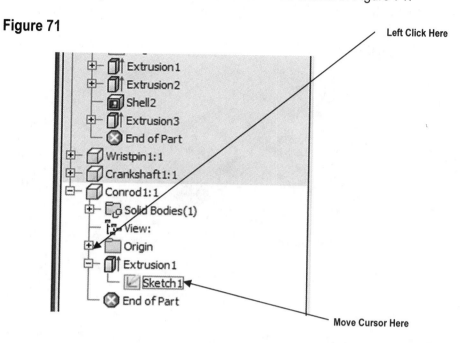

Left Click Here

Move Cursor Here

77. Move the cursor over the text "Sketch1" causing a red box to appear around the text. Notice at the same time the sketch will appear in red on the connecting rod as shown in Figure 72.

Figure 72

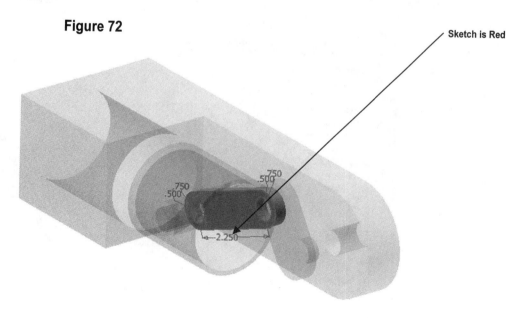

Sketch is Red

78. Right click on **Sketch1** while the red box is visible around the text. A pop up menu will appear. Left click on **Edit Sketch** as shown in Figure 73.

Figure 73

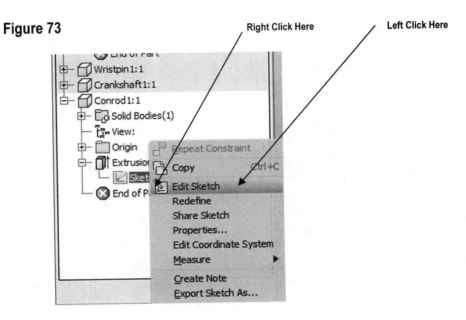

79. Your screen should look similar to Figure 74.

Figure 74

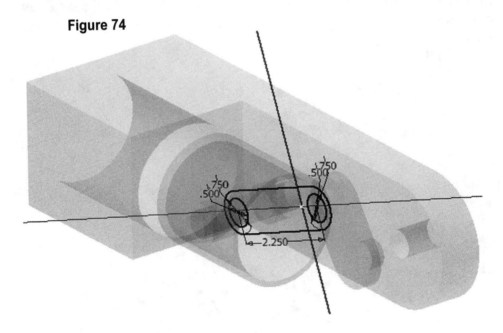

80.   Move the cursor over the 2.25 dimension. After it turns red, double click the left mouse button. The Edit Dimension dialog box will appear as shown in Figure 75.

Figure 75

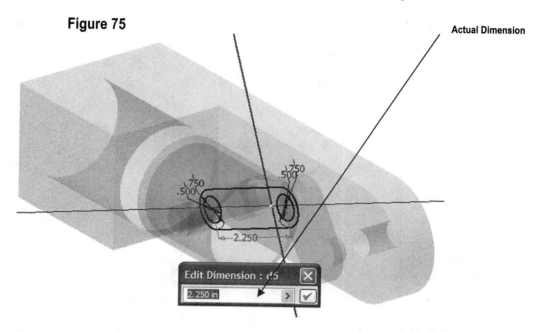

81.   While the text is still highlighted, enter **4.75** as shown in Figure 76 and press **Enter** on the keyboard.

Figure 76

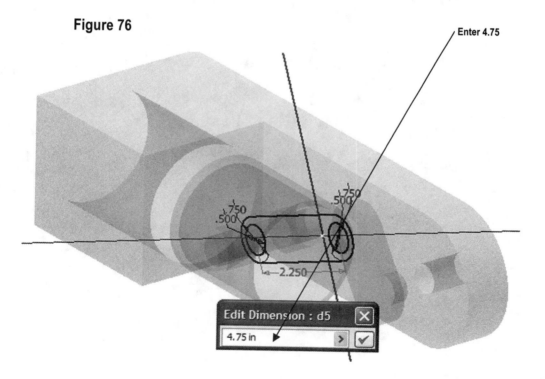

82. The length of the connecting rod will become 4.75 inches as shown in Figure 77.

Figure 77

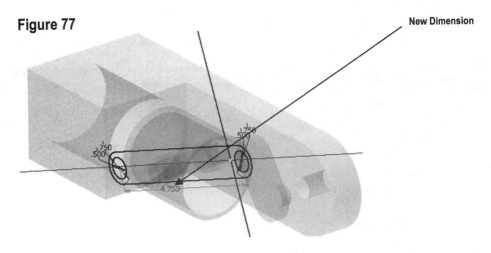

83. Move the cursor to the upper middle portion of the screen and left click the **Manage** tab. Left click on **Update** as shown in Figure 78.

Figure 78

84. Inventor will update the change made to the sketch in the Part Features Panel as shown in Figure 79.

Figure 79

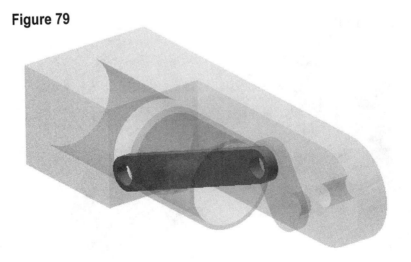

85. Move the cursor to the upper right portion of the screen and left click on **Return** as shown in Figure 80.

Figure 80

Left Click Here

86. Inventor will return to the Assembly Panel displaying the changes made to the connecting rod. Your screen should look similar to Figure 81.

Figure 81

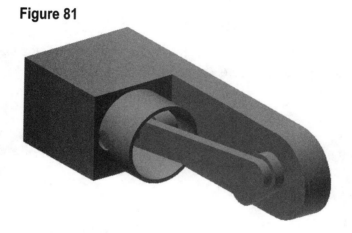

87. The length of the crankshaft pin also must be modified. Move the cursor over the crankshaft as shown in Figure 82. The edges will turn red.

Figure 82

Move Cursor Here

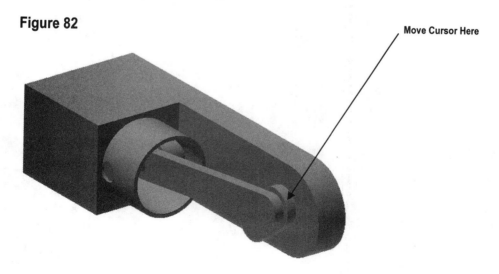

88. Double click (left click) on the crankshaft. All other parts will become grayed as shown in Figure 83.

Figure 83

Double Click Here

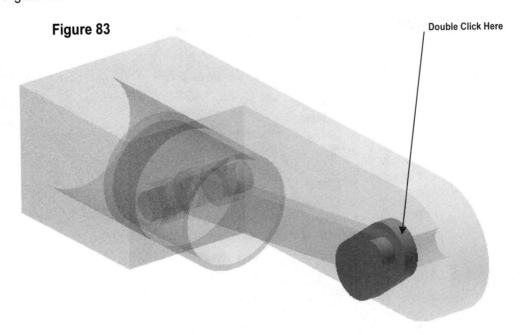

89. Notice that the part tree at the lower left of the screen has changed. All of the branches related to all other parts are grayed (inactive). The branches that illustrate the crankshaft are white (active) as shown in Figure 84.

Figure 84

Inactive Branches   Active Branches

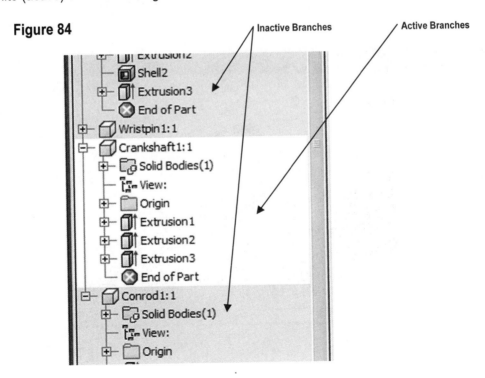

90. Right click on **Extrusion3**. A pop up menu will appear. Left click on **Edit Feature** as shown in Figure 85.

Figure 85

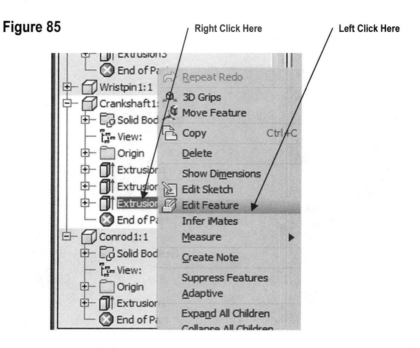

91. The Extrude dialog box will appear. Enter **2.00** for the extrusion distance and left click on **OK** as shown in Figure 86

Figure 86

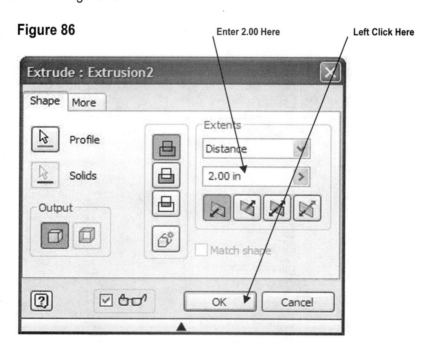

92. Inventor will update the change made to the sketch in the Part Features Panel as shown in Figure 87.

Figure 87

93. Move the cursor to the upper left portion of the screen and left click on the **Manage** tab. Left click on **Update** as shown in Figure 88.

Figure 88

94. Move the cursor to the upper right portion of the screen and left click **Return** as shown in Figure 89.

Figure 89

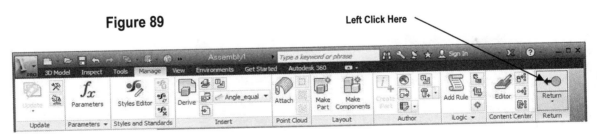

95. Inventor will return to the Assembly Panel displaying the changes made to the crankshaft. Your screen should look similar to Figure 90.

Figure 90

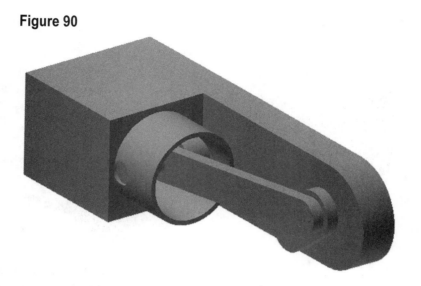

Learn to assign colors to different parts in the Assemble Panel

96. Move the cursor to the upper middle portion of the screen and left click on the double arrows. Left click on the drop down arrow at the right. A drop down menu will appear. Left click on **Materials** as shown in Figure 91.

Figure 91

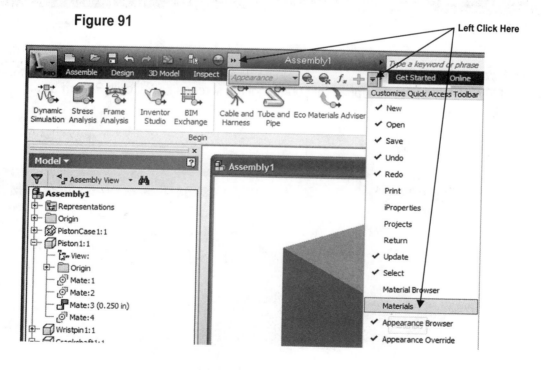

97.   Move the cursor to the left portion of the Piston Case and left click once as shown in Figure 92.

Figure 92

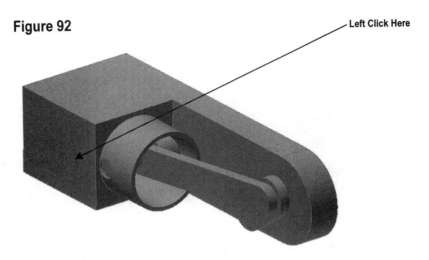

98.   Move the cursor to the upper right portion of the screen and left click on the drop down arrow next to the text "PAEK Plastic". A drop down menu will appear. Scroll down to **Polycarbonate, Clear** and left click once as shown in Figure 93.

Figure 93

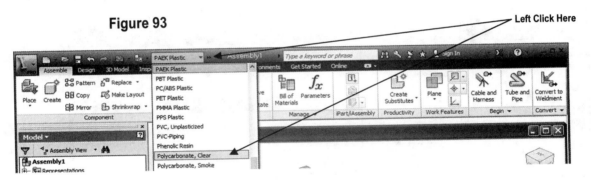

99.   Inventor will change the color of the piston case to Polycarbonate, Clear as shown in Figure 94.

Figure 94

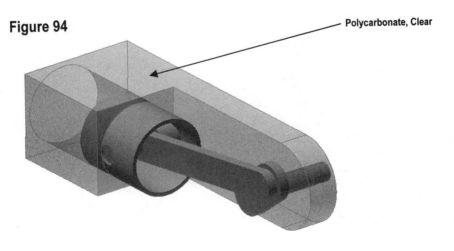

100. Move the cursor to any portion of the piston causing the edges to turn red and left click as shown in Figure 95.

Figure 95

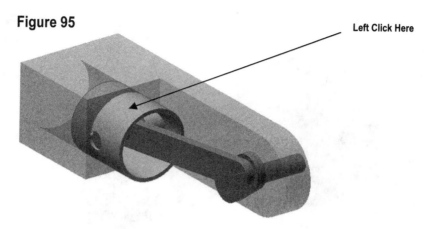

Left Click Here

101. Move the cursor to the upper right portion of the screen and left click on the drop down arrow next to the text "Polycarbonate, Clear". A drop down menu will appear. Scroll down to Stainless Steel and left click as shown in Figure 96.

Figure 96

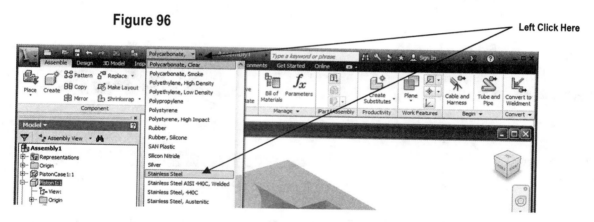

Left Click Here

102. Inventor will change the color of the piston to clear polished blue as shown in Figure 97.

Figure 97

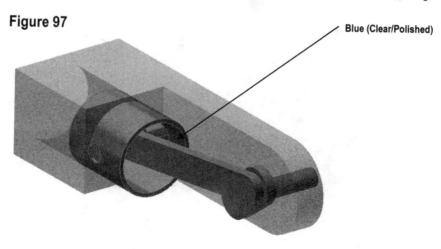

Blue (Clear/Polished)

103. Using the same procedure, change the connecting rod color to **SAN Plastic** as shown in Figure 98.

Figure 98

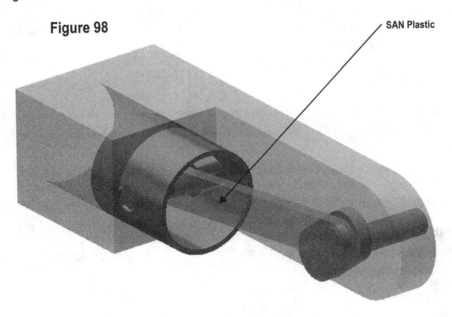

SAN Plastic

104. Move the cursor to the face of the connecting rod causing the edges to turn red. After the edges turn red, left click (holding the left mouse button down) and drag the cursor in a circle causing the crankshaft to turn. Rotate the crankshaft upward to the position shown in Figure 99.

Figure 99

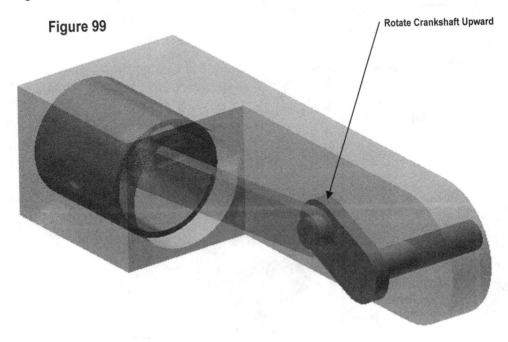

Rotate Crankshaft Upward

Learn to drive constraints to simulate motion

105. Move the cursor to the upper left portion of the screen and left click on the **Assemble** tab. Left click on **Constraint.** The Place Constraint dialog box will appear. Left click on the "Angle Constraint" icon as shown in Figure 100.

Figure 100

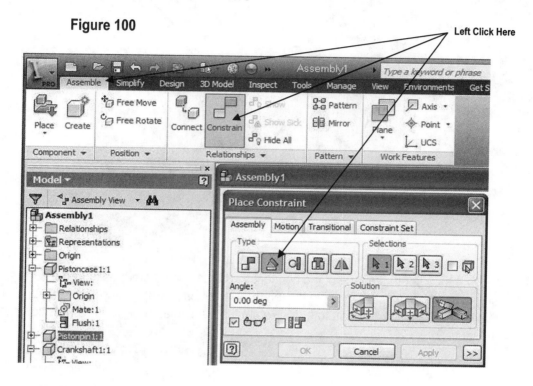

106. Move the cursor to the top portion of the crankshaft causing a red arrow to appear. Left click as shown in Figure 101. You may have to zoom in to select the surface.

Figure 101

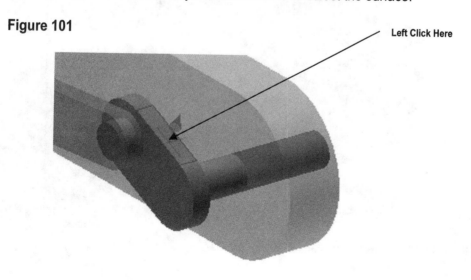

107. Move the cursor to the side of the piston case causing a red arrow to appear. Left click once as shown in Figure 102.

Figure 102

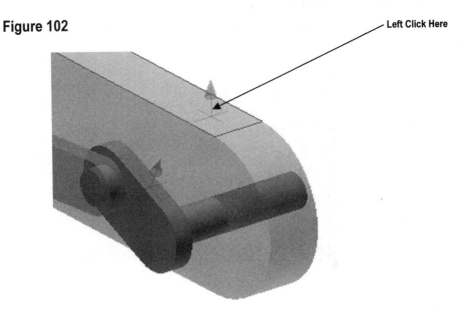

108. Inventor will rotate the crankshaft so that it is parallel (0 degrees) to the side of the piston case. If 20 or 30 degrees were entered in the Angle box, Inventor would rotate the crankshaft to a position 20 or 30 degrees from the side of the piston case. When 0 is entered into the Angle box, Inventor will rotate the crankshaft parallel to the piston case side as shown in Figure 103.

Figure 103

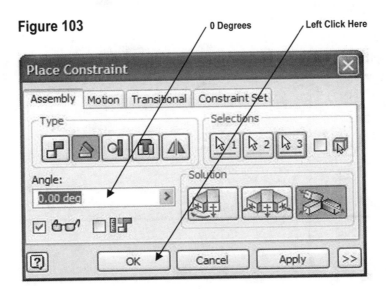

109. Left click on **OK** as shown in Figure 103.

110. Move the cursor to the lower left portion of the screen to the part tree. Scroll down to **Angle:1** and right click once. A pop up menu will appear. Left click on **Drive** or **Drive Constraint** as shown in Figure 104.

Figure 104

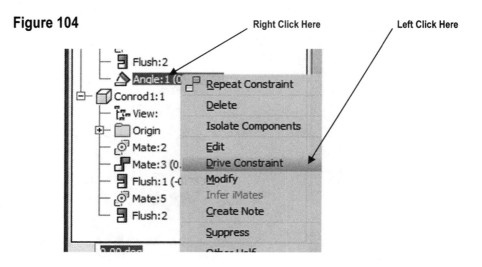

111. The Drive Constraint dialog box will appear. Enter **0** degrees under "Start". Enter **360000** degrees under "End". Left click on the double arrows at the far right lower corner of the dialog box as shown in Figure 105.

Figure 105

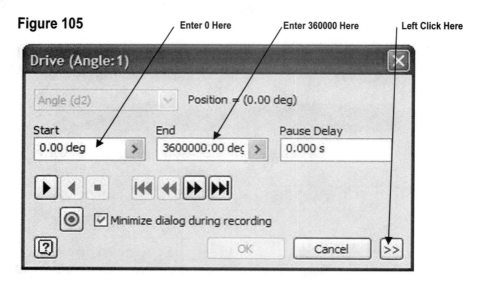

112. The Drive Constraint dialog box will expand, providing more options. Enter **10** for number of degrees as shown in Figure 106.

Figure 106

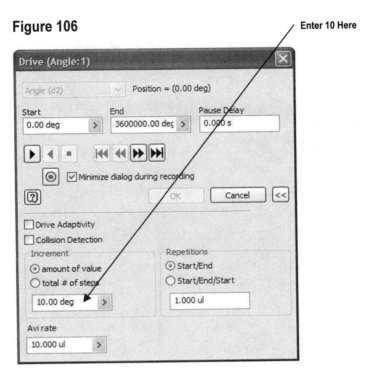

113. Use the Zoom option to zoom out. Use the Pan option to move the assembly off to the side. Left click on the "Play" icon as shown in Figure 107.

Figure 107

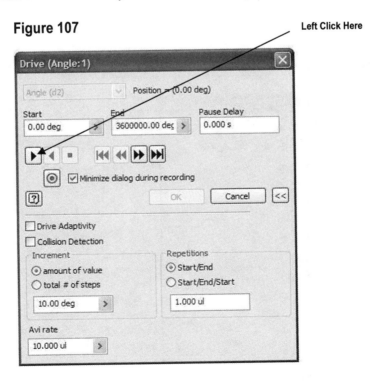

114. Inventor will animate the part causing the crankshaft to rotate.

115. Left click on the "Stop" icon. The animation will stop. Left click on the "Minimize" icon. The Drive Constraint dialog box will get smaller. Left click on the "Rewind" icon. This will rewind the animation back to 0 degrees as shown in Figure 108.

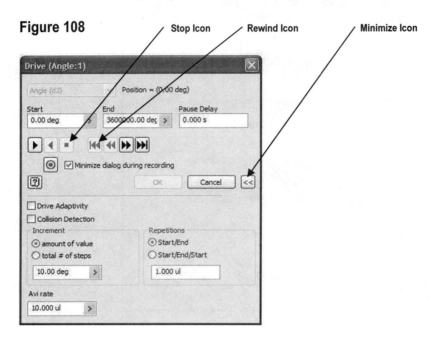

Figure 108

Stop Icon   Rewind Icon   Minimize Icon

Learn to create an .avi or .wmv file while in the Assemble Panel

116. Left click on the "Play" icon and immediately left click on the "Record" icon as shown in Figure 109.

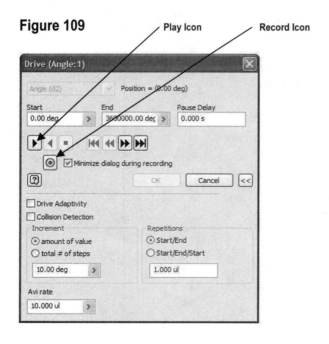

Figure 109

Play Icon   Record Icon

117. The Save As dialog box will appear. Save the file where it can be easily retrieved later and left click on **OK**.

118. The ASF Export Properties dialog box will appear. Left click on **OK** as shown in Figure 110.

Figure 110

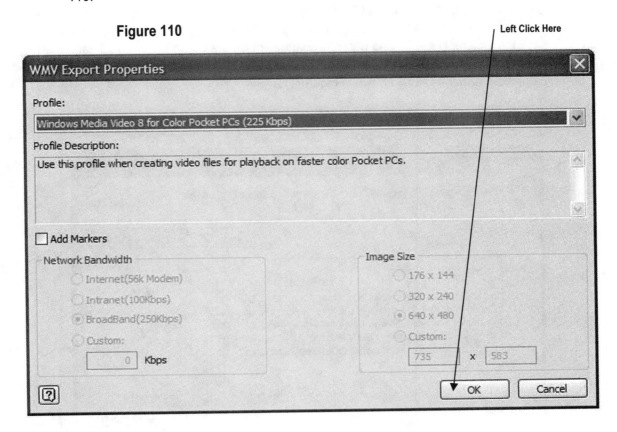

119. While Inventor is recording the simulation, the Drive Constraint dialog box will minimize in the lower left corner of the screen. The speed of the animation will decrease during the recording time. Allow Inventor to record for approximately 15-30 seconds. Inventor is in the process of creating an .wmv file that can be viewed in Windows Media Player. After about 30 seconds left click on the Close symbol in the upper right corner of the dialog box as shown in Figure 111. The Drive Constraint dialog box will close and the recording will be complete.

Figure 111

120. Go to the location where the file was saved and double click on it.

121. Windows Media Player or Real Player will play the file. The file can also be opened in either Windows Media Player or Real Player.

122. Save the Inventor file (.iam) as Chapter 7 Assembly1.iam where it can be easily retrieved at a later time. The Save dialog box will appear. The dialog box will ask if you want to save the assembly itself along with any changes that were made to individual parts that make up the assembly. You can elect to save or not save changes made to individual parts. Left click on **OK** as shown in Figure 112.

Figure 112

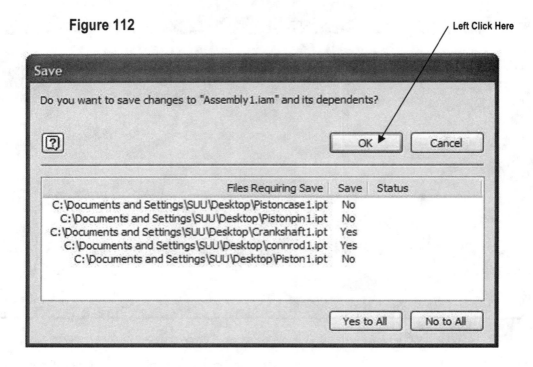

Chapter Problems

Using the dimensions from Chapter 6, modify the PistonCase housing to accept 2 pistons. The length of the shorter crankshaft pin will need to be increased by .25 inches to .050 inches in order to accommodate the additional connection rod as shown.

Start by opening the original Pistoncase1 part file and editing the housing. Then open the original Crankshaft1 part file and increase the extrusion distance of the shorter crankshaft pin to 0.50 inches.

Problem 7-1

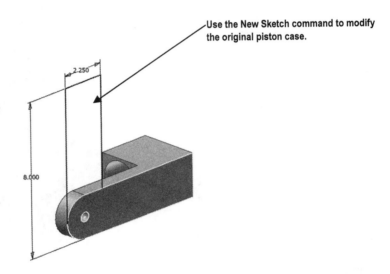

Use the New Sketch command to modify the original piston case.

Problem 7-2

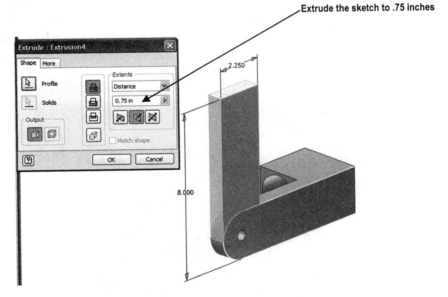

Extrude the sketch to .75 inches

Problem 7-3   Use the Rectangle command to create a new sketch and extrude
                      it to 2.750 inches. Then complete the sketch shown below.

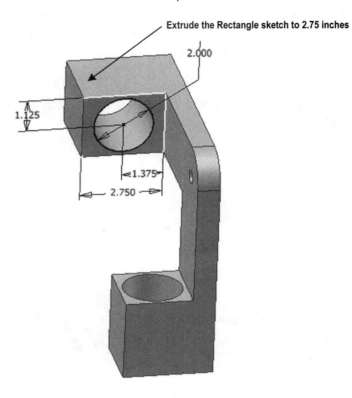

Problem 7-4   Increase the Extrusion distance on one crankshaft pin from .25 to 2.00
                      inches (if this was not already completed from Chapter 7) Increase the
                      extrusion distance from .25 inches to .50 inches on the other pin as
                      shown.

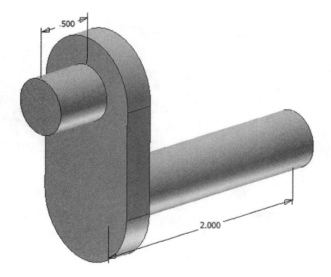

Problem 7-5   Using the Place command, import the parts shown below into the Assembly Panel. You will need 2 pistons, 2 connecting rods and 2 wrist pins, 1 piston case and 1 crankshaft.

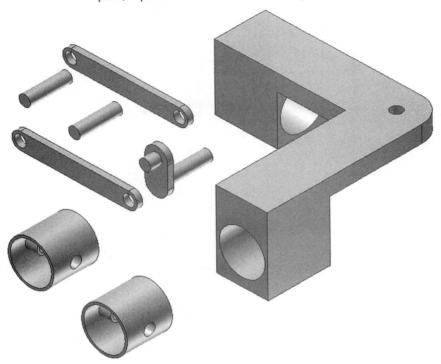

Problem 7-6   Using the dimensions from Chapter 6, assemble all the parts in the same manner as Chapter 6 into a new assembly. When finished your screen should look similar to what is shown below.

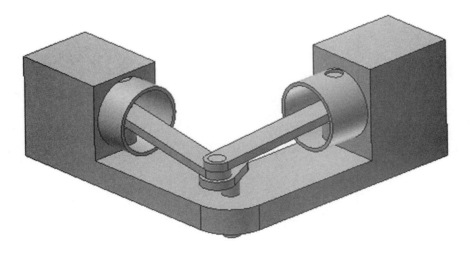

Notes:

Introduction to the Presentation Panel

Objectives:

1. Learn to import existing solid models into the Presentation Panel
2. Learn to design parts trails in the Presentation Panel

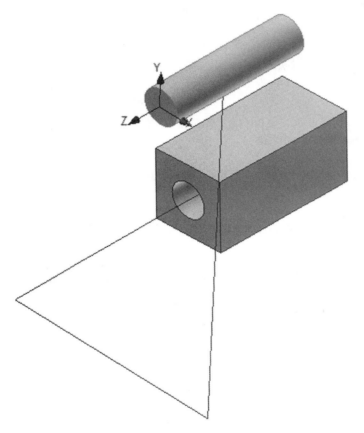

Chapter 8 includes instruction on how to design the presentation shown.

1. Start Inventor 2014 by referring to "Chapter 1 Getting Started".

2. After Inventor is running, begin by creating the parts shown in Figure 1 (pin diameter is .500 inches).

Figure 1

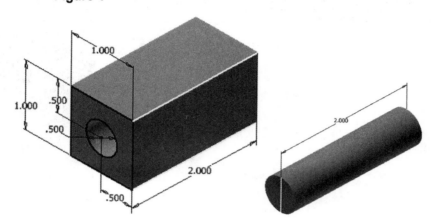

3. Save the block as Chapter 8 Part1.ipt. and save the pin as Chapter 8 Part2.ipt where they can easily be retrieved at a later time. Close both files.

4. Move the cursor to the upper left portion of the screen and left click on the "New" icon as shown in Figure 2.

Figure 2

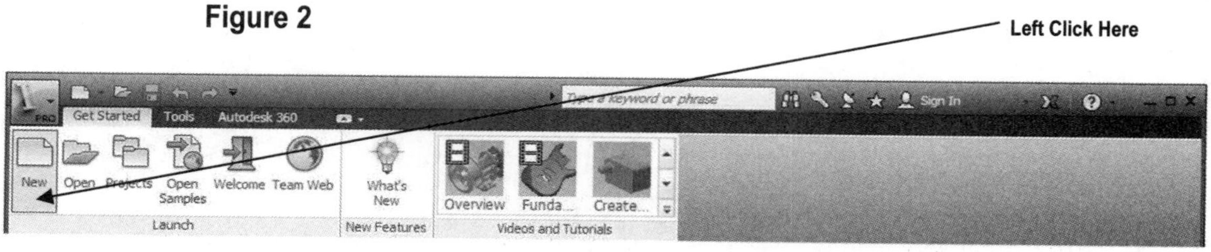

Left Click Here

5.   The Create New File dialog box will appear. Select the **English** folder and **Standard (in).iam** and left click on **OK** as shown in Figure 3.

Figure 3

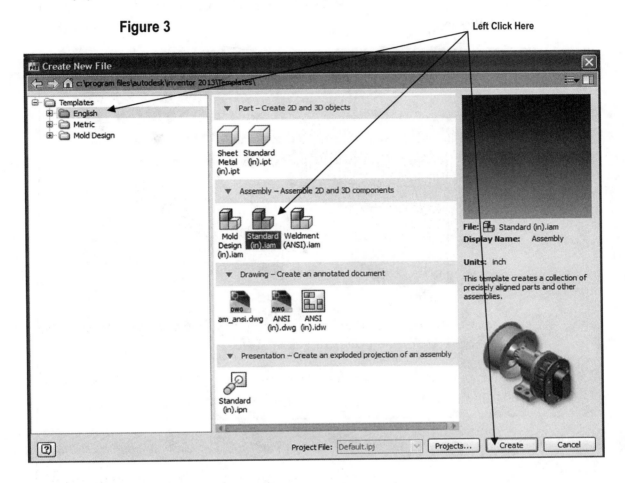

6.   The Assemble Panel will open.

7.   Your screen should look similar to Figure 4. If the Assemble Panel tools are not visible, left click on the **Assemble** tab as shown in Figure 4.

Figure 4

8. Move the cursor to the upper left portion of the screen and left click on **Place** as shown in Figure 5.

Figure 5

Left Click Here

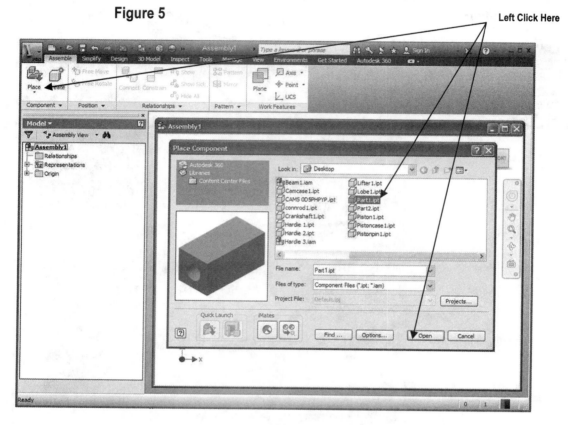

9. The Open dialog box will appear. Left click on **Part1.ipt**. Left click on **Open** as shown in Figure 5.

10. The block will appear attached to the cursor. Do NOT left click. Press **Esc** on the keyboard. Your screen should look similar to Figure 6.

Figure 6

11. Move the cursor to the upper left portion of the screen and left click on **Place** as shown in Figure 7.

Figure 7

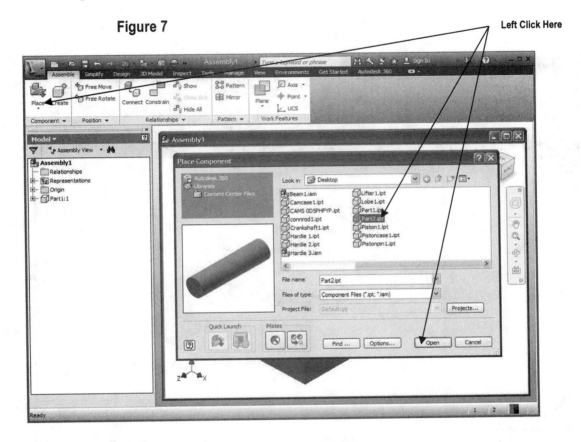

12. The Place Component dialog box will appear. Left click on **Chapter 8 Part2.ipt**. Left click on **Open** as shown in Figure 7.

13. The pin will appear attached to the cursor. Place the pin near the block and left click once. Press **Esc** on the keyboard. Your screen should look similar to Figure 8.

Figure 8

14. Move the cursor to the middle left portion of the screen and left click on **Constrain** as shown in Figure 9.

Figure 9

Left Click Here

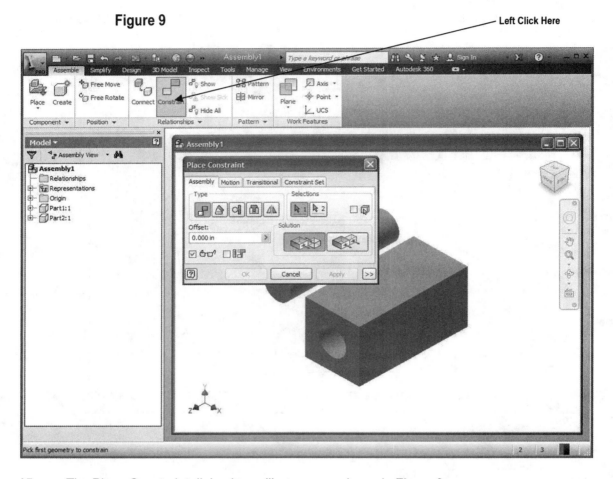

15. The Place Constraint dialog box will appear as shown in Figure 9.

16. Move the cursor over the hole in the block. A red dashed center line will appear. Left click once as shown in Figure 10.

Figure 10

Left Click Here

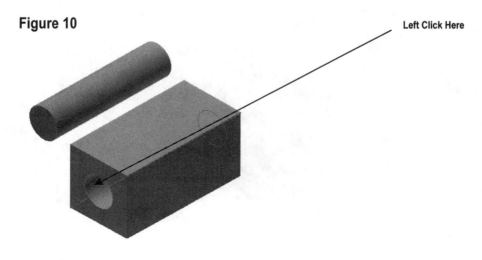

17. Move the cursor over the pin. A red dashed center line will appear. Left click once as shown in Figure 11.

Figure 11

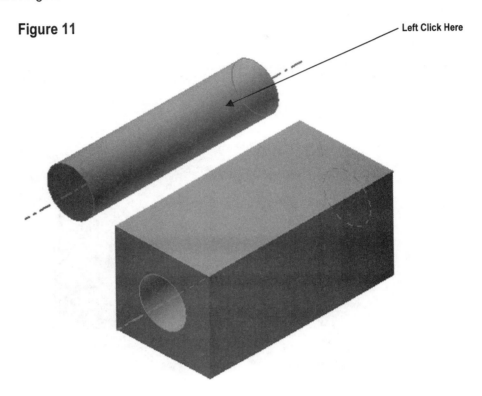

Left Click Here

18. Inventor will insert the pin into the block. Your screen should look similar to Figure 12.

Figure 12

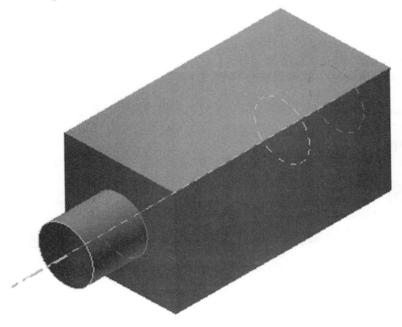

19. Typically a surface constraint would be added to prevent the pin from sliding back and forth in the block. However, this assembly will be used in the Presentation Panel. A surface constraint will not be added because the pin must slide in and out of the block.

20. Move the cursor to the center of the pin. Left click (holding down the left mouse button) and slide the pin flush with the outside of the block as shown in Figure 13.

Figure 13

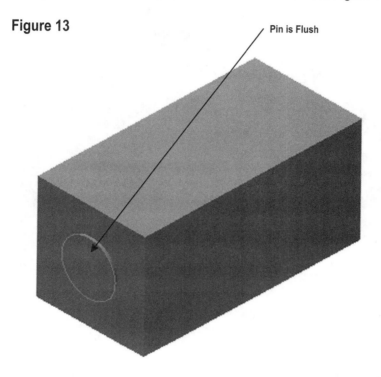

Pin is Flush

21. Save the parts as Chapter 8 Assembly1.iam where it can be easily retrieved later. Leave the file open at this time.

22. Move the cursor to the upper left portion of the screen and left click on the "New" icon. Left click on **Presentation** as shown in Figure 14.

Figure 14

Left Click Here

Learn to import existing assembly models into the Presentation Panel

23. The New File dialog box will appear. Select the **English** tab and **Standard (in).ipn**. Left click on **OK** as shown in Figure 15.

Figure 15

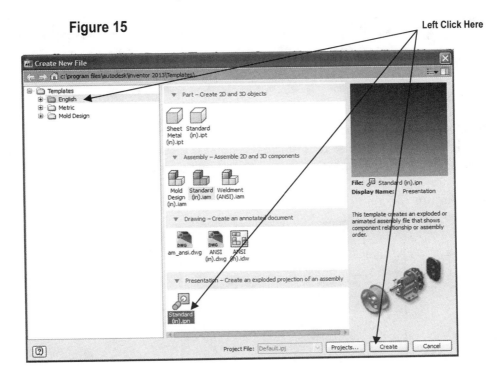

24. The Presentation Panel will open as shown in Figure 16.

Figure 16

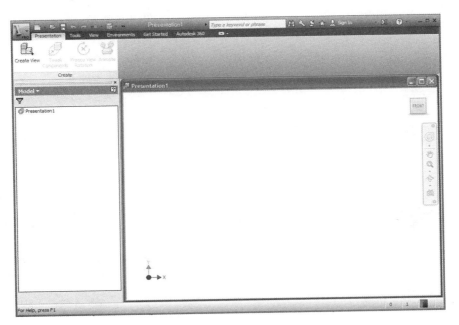

25. Move the cursor to the upper left portion of the screen and left click on **Create View** as shown in Figure 17.

Figure 17

Left Click Here

26. The Select Assembly dialog box will appear. Left click on the "Explore" icon located at the upper right portion of the dialog box if the assembly does not appear. Left click on **OK** as shown in Figure 18. If an error message appears, left click on OK.

Figure 18

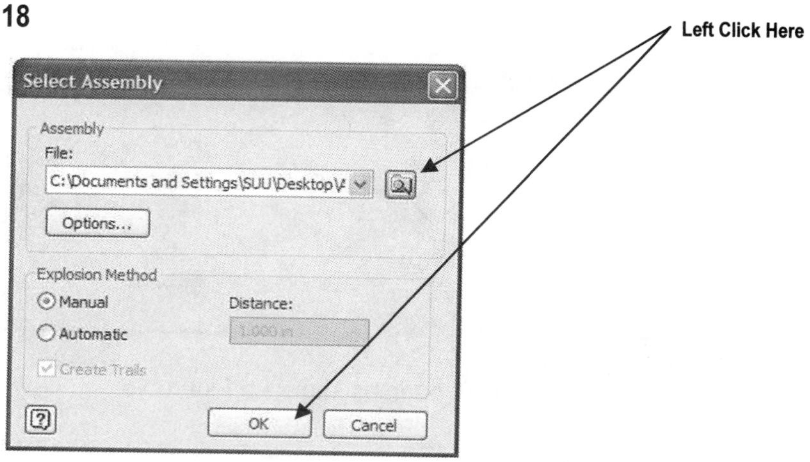

Left Click Here

27. Left click on **Assembly1.iam**. Left click on **Open** as shown in Figure 19.

Figure 19

Left Click Here

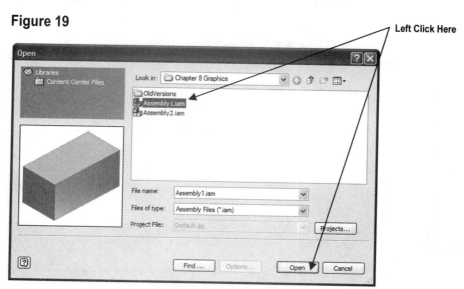

28. *The Presentation Panel will only read assembly drawings.* Assembly drawings are imported into the Presentation Panel in order to create an .ipn file (Inventor Presentation).

29. The Select Assembly dialog box will open. Left click on **OK** as shown in Figure 20.

Figure 20

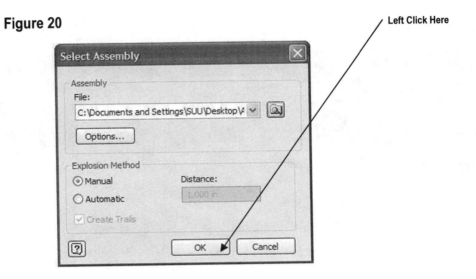

30. Your screen should look similar to Figure 21.

Figure 21

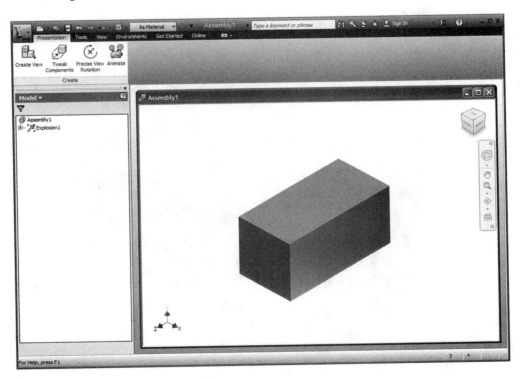

Learn to design parts trails in the Presentation Panel

31. Move the cursor to the upper left portion of the screen and left click on **Tweak Components**. Left click on the "Z" icon as shown in Figure 22.

Figure 22

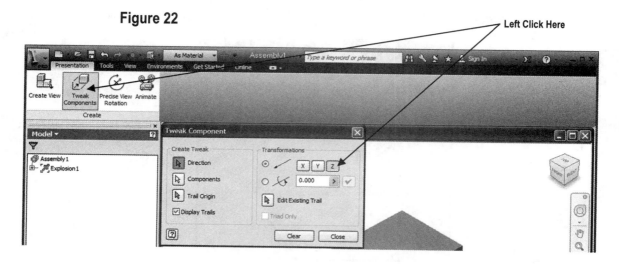

32. Move the cursor to the face of the pin. An origin symbol will be attached to the cursor. After the origin symbol appears, left click once as shown in Figure 23.

Figure 23

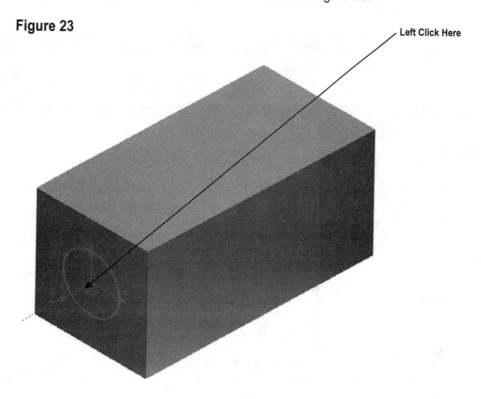

33. Move the cursor to the face of the pin. Left click (holding the left mouse button down) and drag the pin out of the block towards the lower left portion of the screen as shown in Figure 24. Notice the blue line coming out of the hole in the block. This is the "trail" that the pin will follow. This is the "Z" axis.

Figure 24

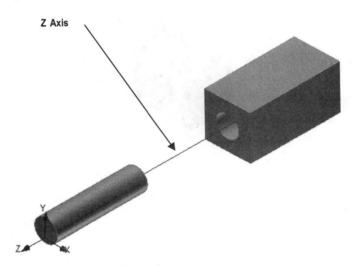

34. Enter **3.000** for the distance that appears below X, Y, Z. This is the distance the pin has traveled on the Z axis. The text can be highlighted so users can enter any numerical distance. Move the cursor to the Tweak Component dialog box and left click on **X** as shown in Figure 25.

Figure 25

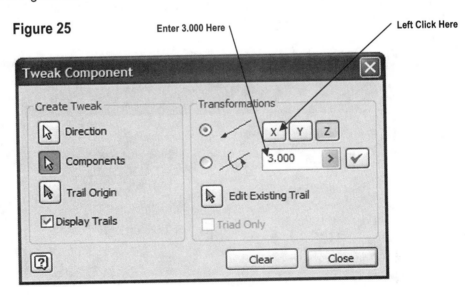

35. Move the cursor to the center of the pin. Left click (holding the left mouse button down) and drag the pin from the end of the "Z" trail, towards the lower right portion of the screen as shown in Figure 26. Notice the direction change of the blue line. This is the "trail" that the pin will follow. This is the "X" axis.

Figure 26

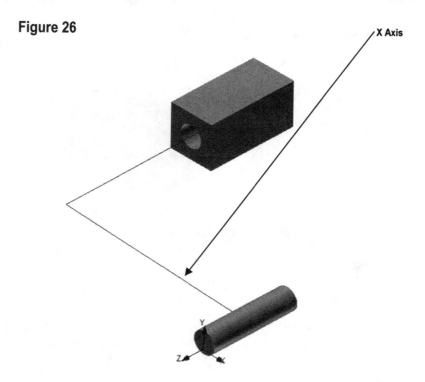

36. Enter **-3.000** for the distance. Move the cursor to the Tweak Component dialog box and left click on **Y** as shown in Figure 27.

Figure 27

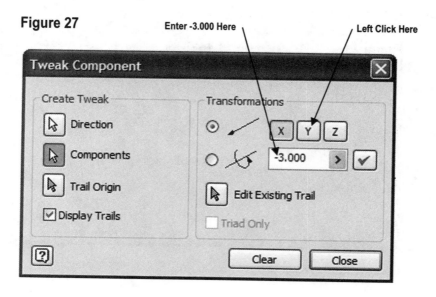

37. Move the cursor to the center of the pin. Left click (holding the left mouse button down) and drag the pin from the end of the "Y" trail upward towards the middle portion of the screen as shown in Figure 28. Notice the direction change of the blue line. This is the "trail" that the pin will follow. This is the "Y" axis.

Figure 28

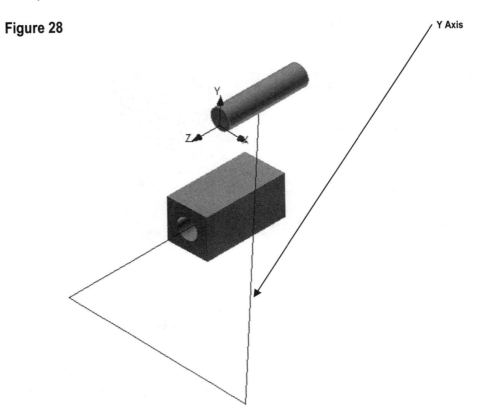

38. Enter 5.**000** for the Distance. Left click on **Clear** and then **Close** as shown in Figure 29.

Figure 29

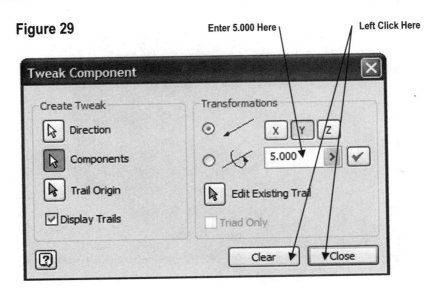

39. The Tweak Component dialog box will close.

40. Move the cursor to the upper left portion of the screen and left click on **Animate.** The Animation dialog box will appear. Left click on "Play" as shown in Figure 30.

Figure 30

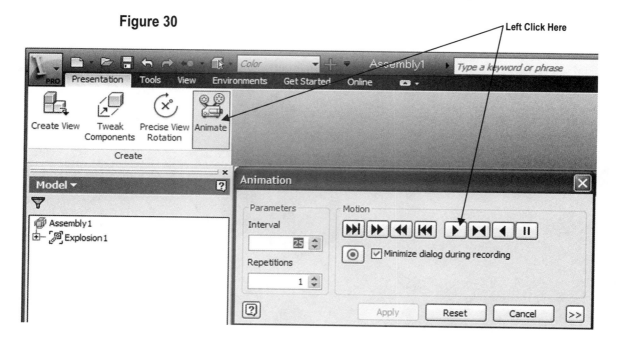

41. Inventor will animate the parts. The pin should follow the part trail back to the hole in the block. Your screen should look similar to Figure 31.

Figure 31

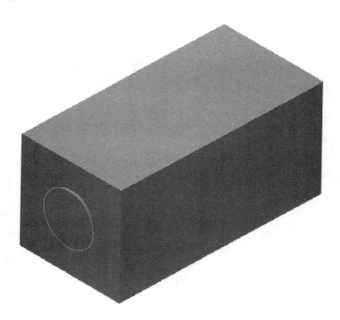

Chapter Problems

Create the parts the following parts and use them to design Inventor Presentations.

Problem 8-1

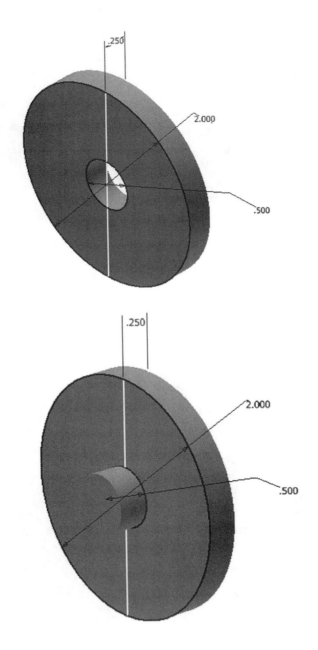

Problem 8-2

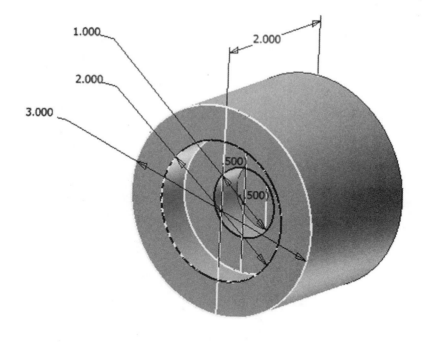

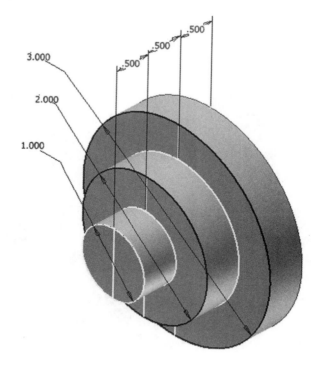

Problem 8-3

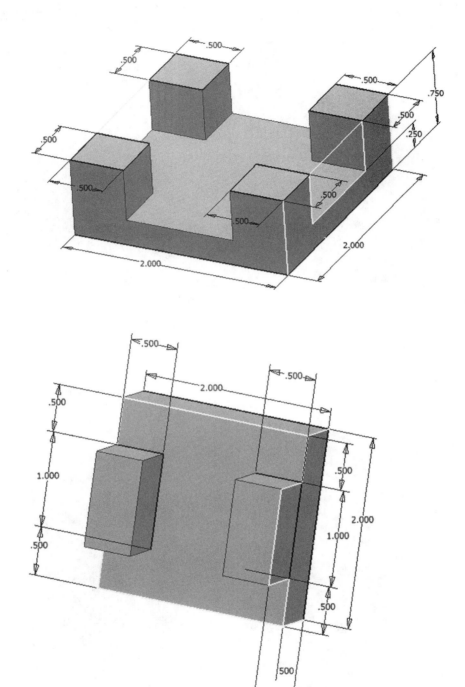

Problem 8-4   Import the following part twice into the **Assembly Panel** and then into the **Presentation Panel**

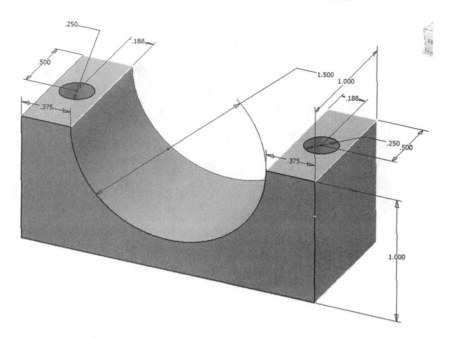

Your screen should similar to what is shown below

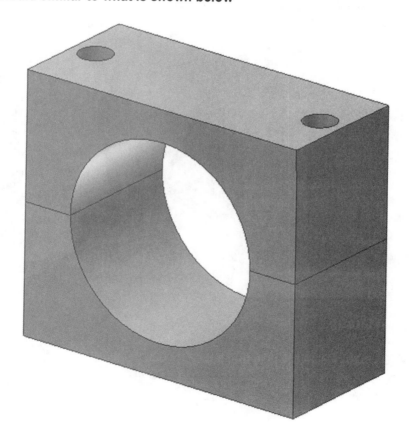

CHAPTER 9

Introduction to Advanced Commands

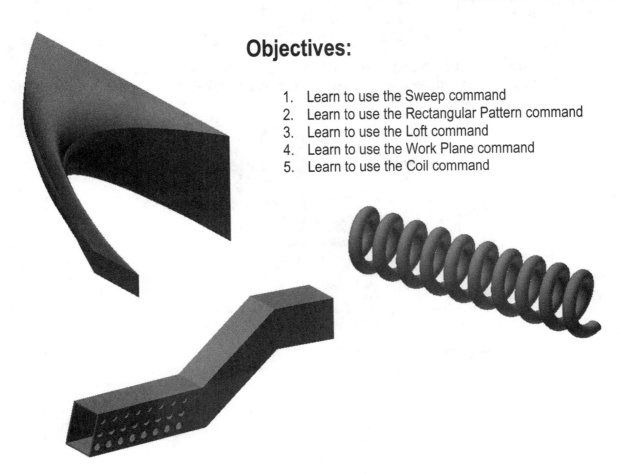

Objectives:

1. Learn to use the Sweep command
2. Learn to use the Rectangular Pattern command
3. Learn to use the Loft command
4. Learn to use the Work Plane command
5. Learn to use the Coil command

Chapter 9 includes instruction on how to design the parts shown.

Learn to create a sweep using the Sweep command

1. Start Inventor by referring to "Chapter 1 Getting Started".

2. After Inventor is running, begin a New Sketch.

3. Move the cursor to the upper left portion of the screen and left click on **Rectangle** as shown in Figure 4.

Figure 1

4. Complete the sketch shown below. Once the sketch is complete, right click anywhere on the screen. A pop up menu will appear. Left click **Finish 2D Sketch** as shown in Figure 2.

Figure 2

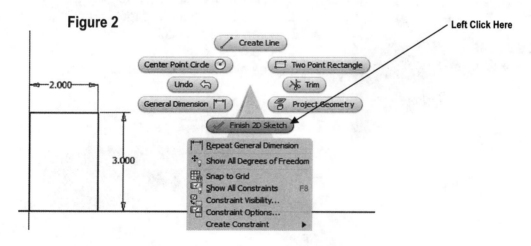

5. Your screen should look similar to Figure 3.

Figure 3

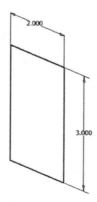

6.   Move the cursor over the text "YZ Plane" causing a red box to appear. Right click once. A pop up menu will appear. Left click on **New Sketch** as shown in Figure 4.

Figure 4

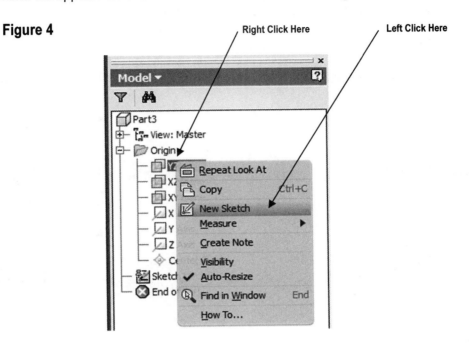

7.   Your screen should look similar to Figure 5.

Figure 5

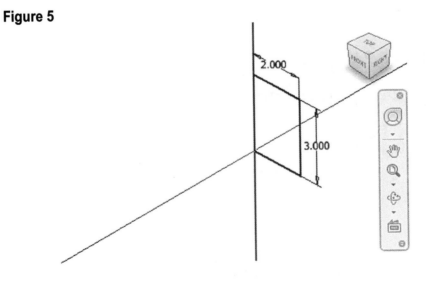

8. Complete the following sketch. The angle of the lines can be estimated. The sketch lines must intersect with the corner of the 2 inch by 3 inch box as shown in Figure 6. Remember to use the **Aligned** dimension function (while the dimension is attached to the cursor right click causing a pop up menu to appear then left click on **Aligned**). Exit out of the Sketch Panel.

Figure 6

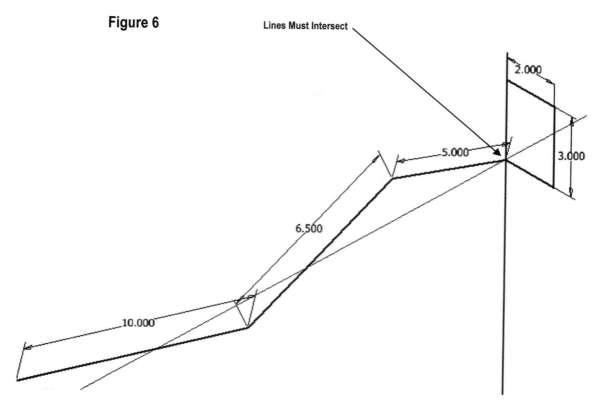

9. Your screen should look similar to Figure 7.

Figure 7

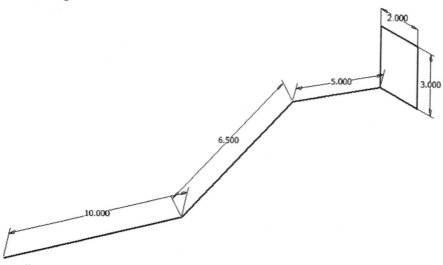

10. Move the cursor to the upper left portion of the screen and left click on **Sweep**. Move the cursor over the sweep line causing it to turn red and left click once as shown in Figure 8.

Figure 8

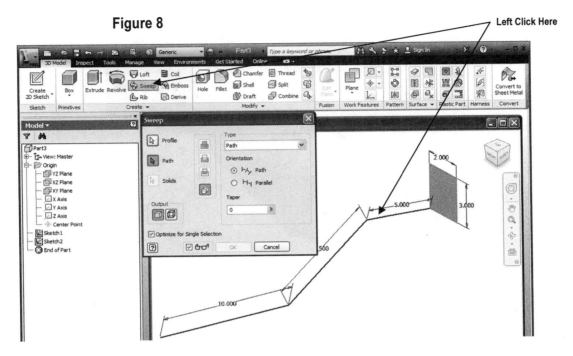

11. A preview of the sweep will appear as shown in Figure 9.

Figure 9

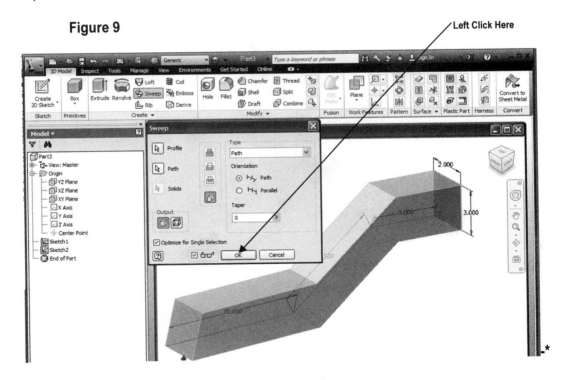

12. Left click on **OK** as shown in Figure 9.

13. Your screen should look similar to Figure 10.

Figure 10

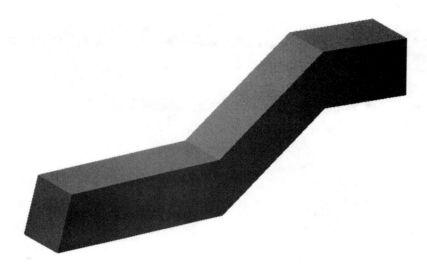

14. Use the **Shell** command to shell the tubing as shown in Figure 11.

Figure 11

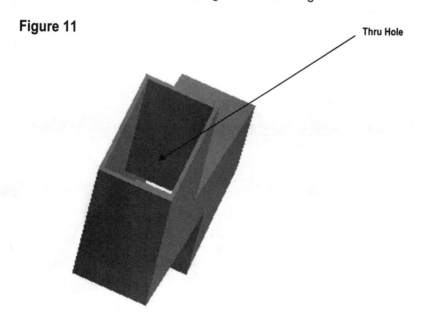

Thru Hole

15. Begin a new sketch as shown in Figure 12. Use the "Look At/View Face" command to gain a perpendicular view of the surface.

Figure 12

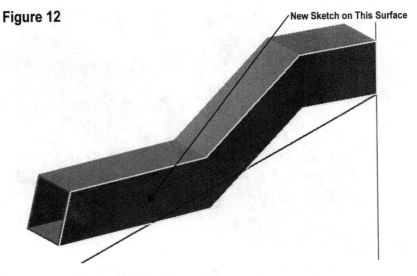

New Sketch on This Surface

16. Your screen should look similar to Figure 13.

Figure 13

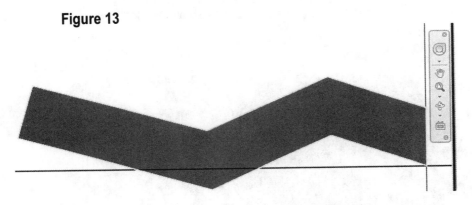

17. Complete the following sketch. You may have to zoom in as shown in Figure 14.

Figure 14

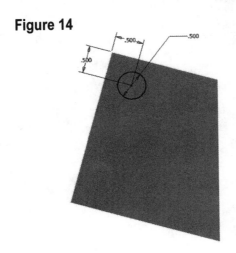

18.   Exit out of the Sketch area and change the view to Home View as shown in Figure 15.

Figure 15

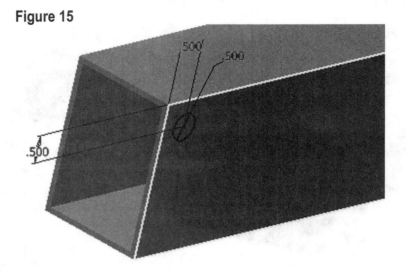

Learn to use the Rectangular Pattern command

19.   Use the **Extrude** command to cut the hole out as shown in Figure 16.

Figure 16

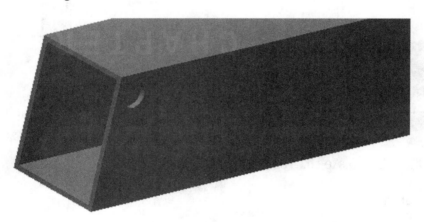

20. Move the cursor to the upper middle portion of the screen and left click on the Rectangular Pattern icon. The Rectangular Pattern dialog box will appear as shown in Figure 17.

Figure 17

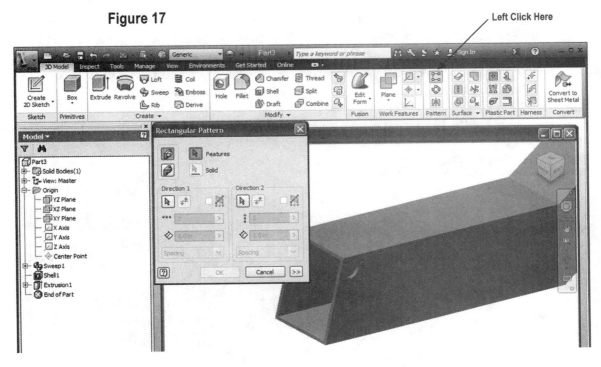

21. Move the cursor inside the hole and left click once as shown in Figure 18.

Figure 18

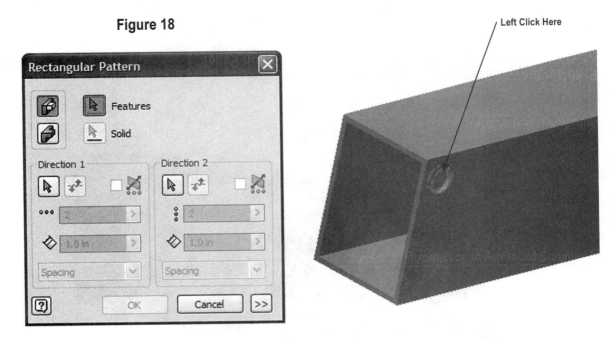

22. Left click on the arrow under Direction 1. Now left click on the top edge of the part. If there is a need to reverse the direction of the pattern, (so the pattern does not travel off the part) left click on the Flip icon (icon with arrows pointing up and down). Inventor will provide a preview of the holes it will pattern in the X direction. Now left click on the arrow under Direction 2. Now left click on the side edge of the part. Enter **8** for the number of occurrences in Direction 1 and **3** for the number of occurrences in Direction 2. Enter **1.0** and **.875** for the distance between the circles respectively. Left click on **OK** as shown in Figure 19.

Figure 19

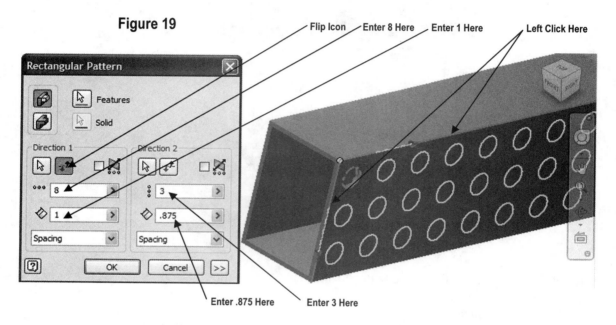

Flip Icon   Enter 8 Here   Enter 1 Here   Left Click Here

Enter .875 Here   Enter 3 Here

Learn to create a loft using the Loft command

23. Your screen should look similar to Figure 20.

Figure 20

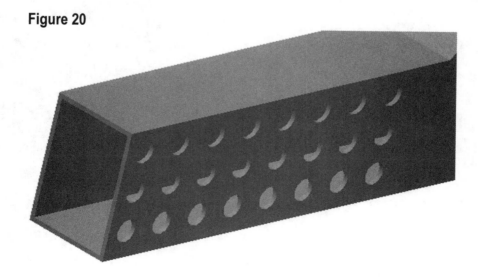

24. Begin a new drawing. Complete the sketch shown below. Exit out of the Sketch Panel as shown in Figure 21.

Figure 21

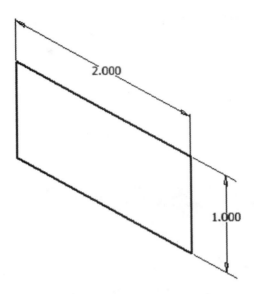

25. Left click on the "plus" sign to the left of the text "Origin". The part tree will expand as shown in Figure 22.

Figure 22

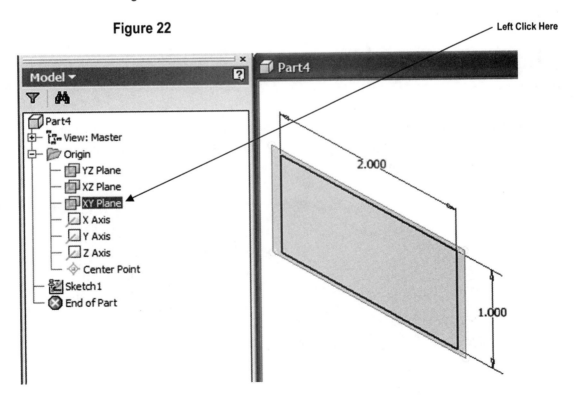

26. Move the cursor to the upper right portion of the screen and left click on **Plane** as shown in Figure 23.

Figure 23

27. Left click on **XY Plane** in the part tree. The "XY Plane" text will become highlighted as shown in Figure 24.

Figure 24

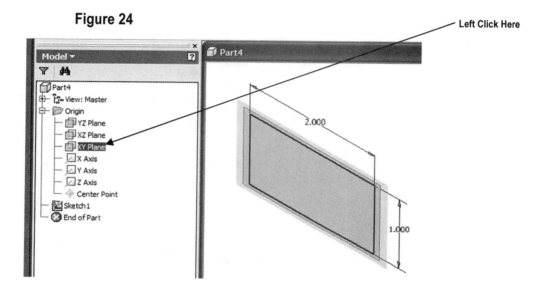

28. Move the cursor to the center of the sketch and left click (holding the left mouse button down) dragging the cursor to the lower left portion of the screen. Enter **.500** as shown in Figure 25 and press the **Enter** key on the keyboard.

Figure 25

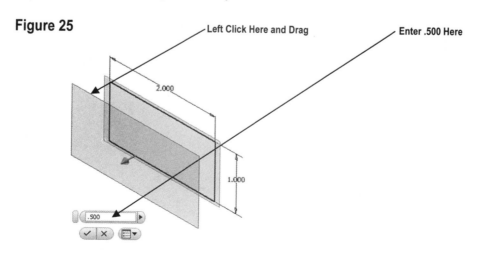

29. Your screen should look similar to Figure 26.

Figure 26

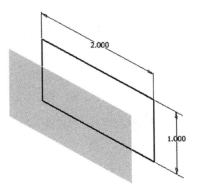

30. Begin a New Sketch on the newly created Plane as shown in Figure 27.

Figure 27

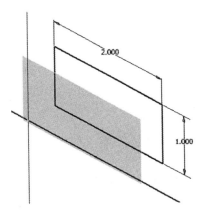

31. Complete the sketch shown in Figure 28. Estimate the location and size of the circle and exit the Sketch Panel.

Figure 28

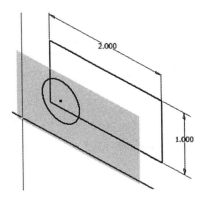

32. Move the cursor to the upper right portion of the screen and left click **Plane** as shown in Figure 29.

Figure 29

Left Click Here

33. Left click on **XY Plane** in the part tree. The "XY Plane" text will become highlighted as shown in Figure 30.

Figure 30

Left Click Here

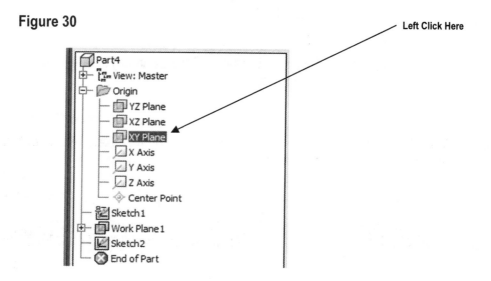

34. Move the cursor to the center of the sketch and left click (holding the left mouse button down) dragging the cursor to the lower left portion of the screen. Enter **1.000** as shown in Figure 31 and press the **Enter** key on the keyboard.

Figure 31

Enter 1.000 Here

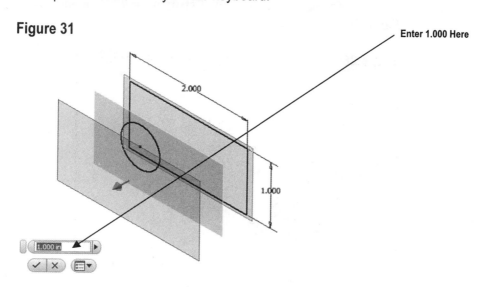

35. Complete the sketch shown in Figure 32. Estimate the location and size of the rectangle and exit the Sketch Panel.

Figure 32

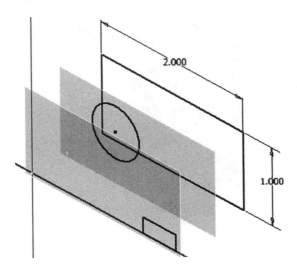

36. Move the cursor to the upper left portion of the screen and left click on **Loft** as shown in Figure 33.

Figure 33

Left Click Here

37. The Loft dialog box will appear as shown in Figure 34.

Figure 34

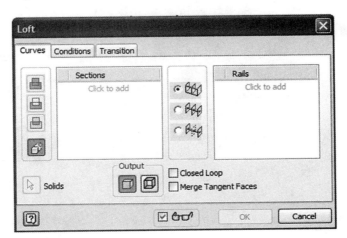

38. Left click on each of the sketches as shown in Figure 35.

Figure 35

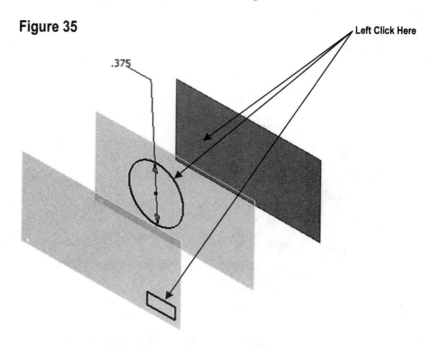

39. Inventor will provide a preview of the loft as shown in Figure 36.

Figure 36

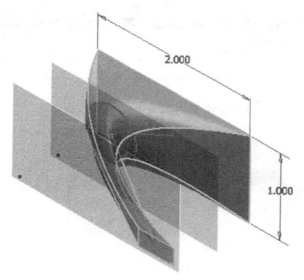

40. Left click on **OK**.

41. Your screen should look similar to Figure 37.

Figure 37

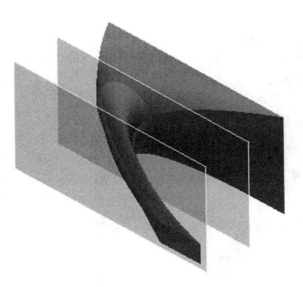

42. To hide the work planes, move the cursor over the edge of the work plane causing the edges to turn red. Right click once. A pop up menu will appear. Left click on **Visibility**. Hide both work planes as shown in Figure 38.

Figure 38

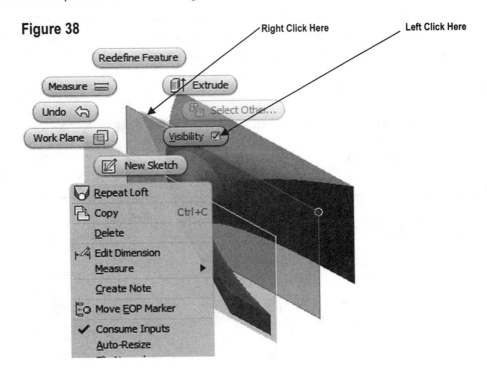

43. Your screen should look similar to Figure 39.

Figure 39

Learn to create a coil using the Coil command

44. If more accuracy is required for each sketch, geometry can be projected from one work plane to another as previously described in Chapter 6. Geometry can also be sized and located using the Dimension command as discussed in previous chapters.

45. Begin a new drawing.

46. Complete the sketch shown in Figure 40 and exit the Sketch Panel.

Figure 40

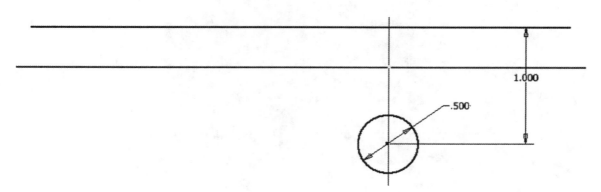

47.   Move the cursor to the middle left portion of the screen and left click on **Coil**. Left click on the horizontal line above the circle as shown in Figure 41.

Figure 41   Left Click Here

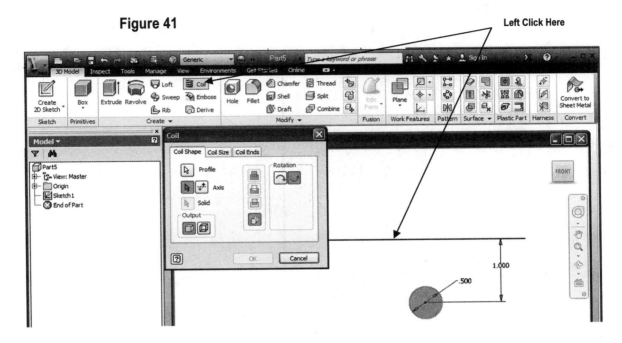

48.   Inventor will provide a preview of the coil. Left click on the Axis icon to reverse the direction of the coil as shown in Figure 42.

Figure 42   Left Click Here

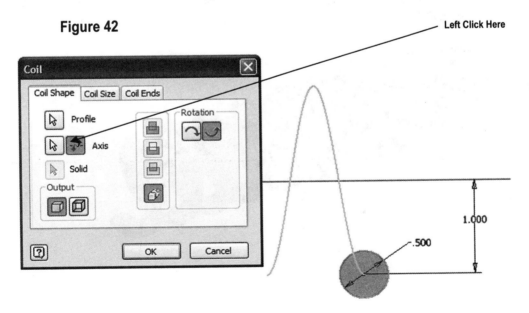

49. Left click on the **Coil Size** tab. Under Revolution enter **10**. Left click on **OK** as shown in Figure 43.

Figure 43

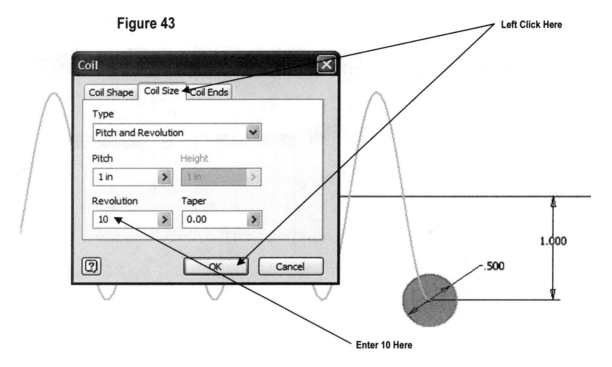

50. Your screen should look similar to Figure 44.

Figure 44

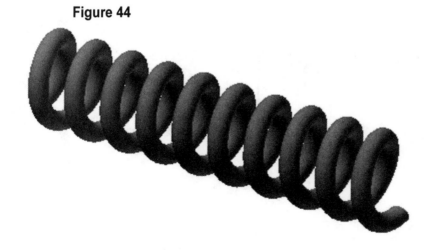

Chapter Problems

Complete the following problem sketches.

Problem 9-1

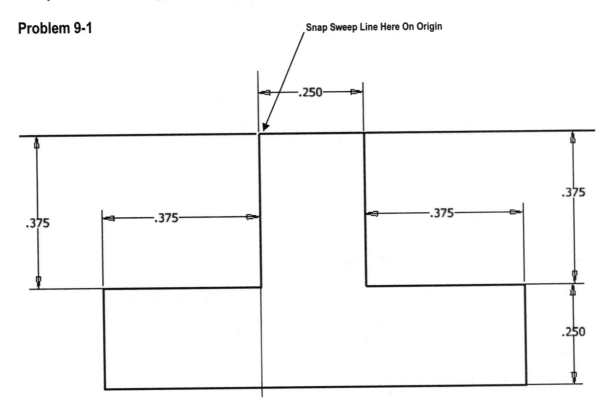

Create the following Sweep line. Sweep the sketch along the sweep line shown below.

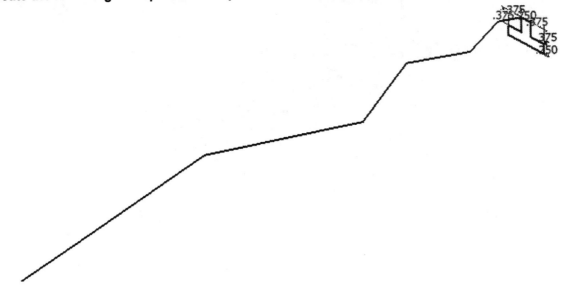

Problem 9-2

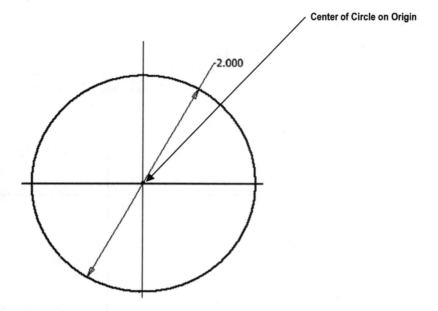

Create the following Sweep line. The center of the circle must intersect the sweep line as shown.

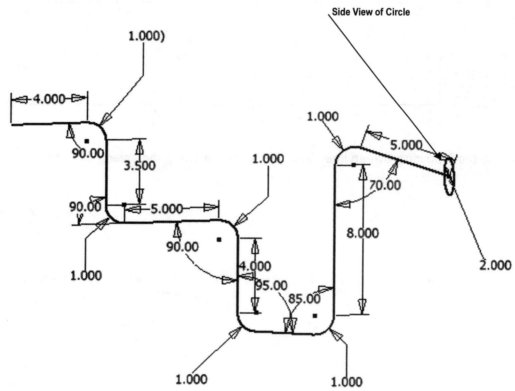

CHAPTER 10

Introduction to Creating Threads

Objectives:

1. Learn to use the Polygon command
2. Learn to create threads in a solid model
3. Learn to interpret threads specs

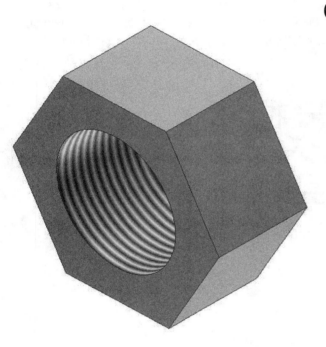

Chapter 10 includes instruction on how to create threads in the part shown.

1.   Start Inventor by referring to "Chapter 1 Getting Started". After Inventor is running, begin a New Sketch.

2.   Move the cursor to the middle left portion of the screen and left click on **Polygon** as shown in Figure 1.

Figure 1   Left Click Here

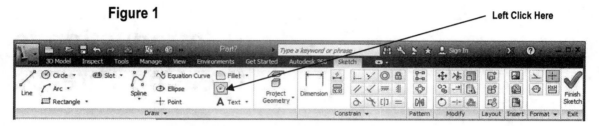

3.   The Polygon dialog box will appear. Left click on the **Circumscribed** icon. This will be used to define the distance across the flats (basically what size wrench or socket would be used to loosed or tighten the nut). Enter **6** for the number of sides as shown in Figure 2.

Figure 2   Left Click Here   Enter 6 Here

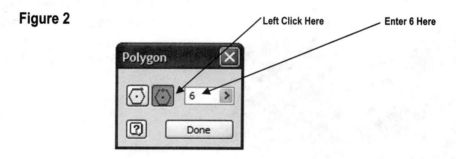

4.   Left click on the origin and drag the cursor out to the left similar to creating a circle as shown in Figure 3.

Figure 3   Left Click Here

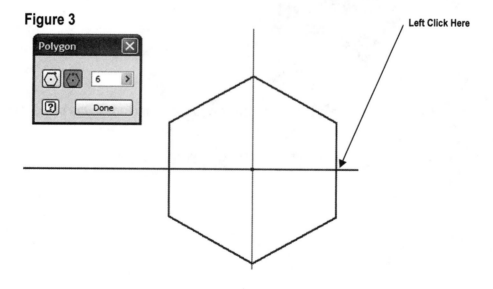

5. Left click on **Done** as shown in Figure 4.

Figure 4

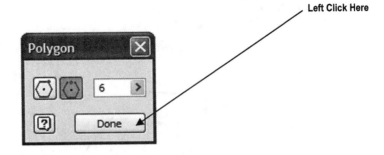

6. Complete and dimension the sketch as shown in Figure 5.

Figure 5

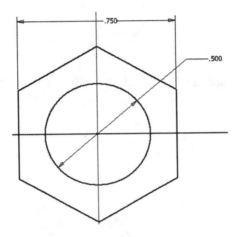

7. Exit out of the Sketch Panel as shown in Figure 6.

Figure 6

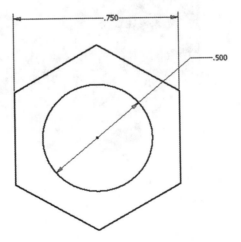

8. Rotate the part around to gain an isometric view of the part as shown in Figure 7.

Figure 7

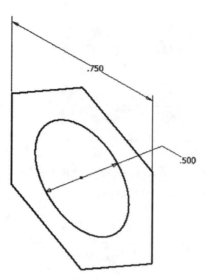

9. Extrude the part a distance of *.375* inches as shown in Figure 8.

Figure 8

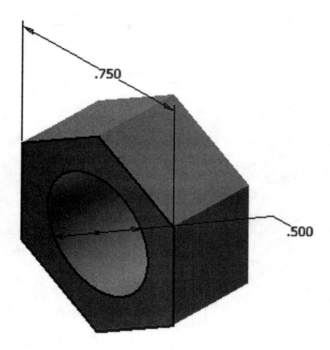

Learn to create Threads

10. Move the cursor to the upper middle portion of the screen and left click on **Thread.** The Thread dialog box will appear as shown in Figure 9.

Figure 9

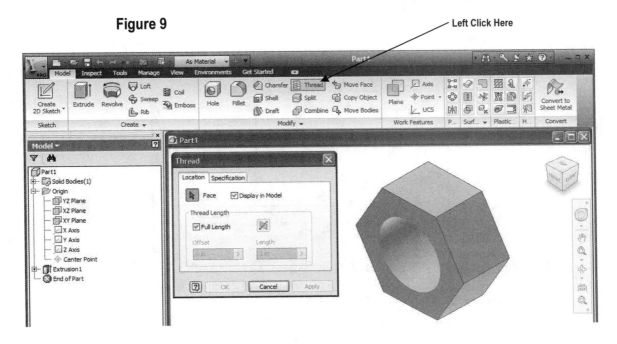

Left Click Here

11. Move the cursor inside the hole and left click once as shown in Figure 10.

Figure 10

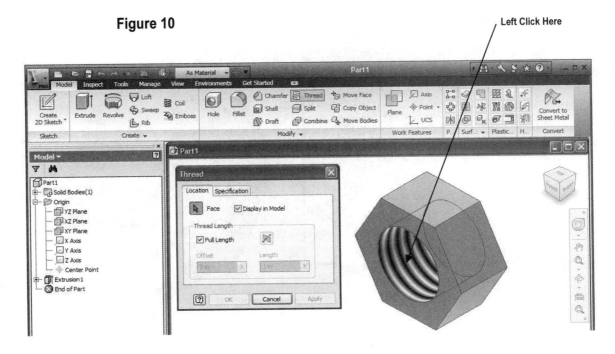

Left Click Here

12. Left click on the **Specification** tab. Left click on the drop down arrow located underneath Designation. A drop down menu will appear. A list of compatible threads will appear. The first number refers to the diameter of the thread (bolt or hole). The second number refers to the pitch (in the case, the number of threads per inch). The letters indicate what type of thread and whether the threads are coarse of fine as shown in Figure 11.

Figure 11

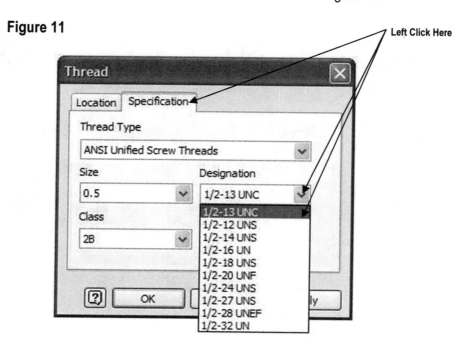

13. Inventor will provide a preview of the different thread types selected as shown in Figure 12.

Figure 12

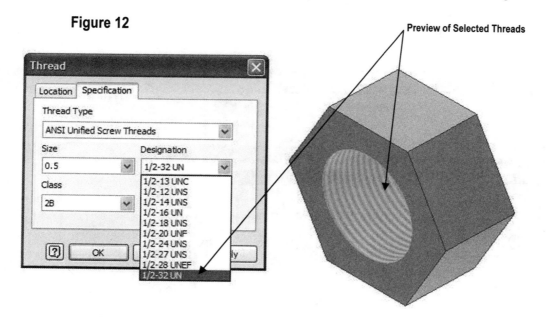

14. Select a thread of your choice and left click on **OK** as shown in Figure 13.

Figure 13

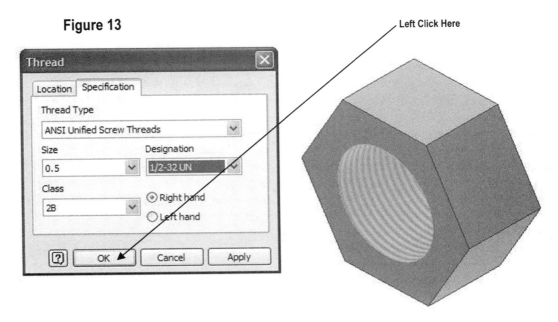

15. Your screen should look similar to Figure 14.

Figure 14

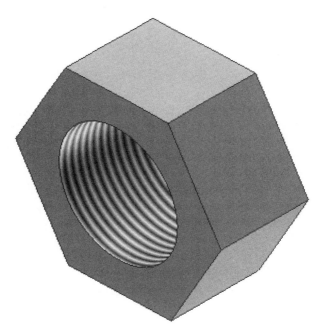

Notes:

CHAPTER 11

Advanced Work Plane Procedures

Objectives:

1. Learn to create points on a solid model
2. Learn to use the Split command
3. Learn to create an offset/oblique Work Plane

Chapter 11 includes instruction on how to design the part shown.

1. Start Inventor by referring to "Chapter 1 Getting Started". After Inventor is running, begin a New Sketch.

2. Complete the sketch shown in Figure 1.

Figure 1

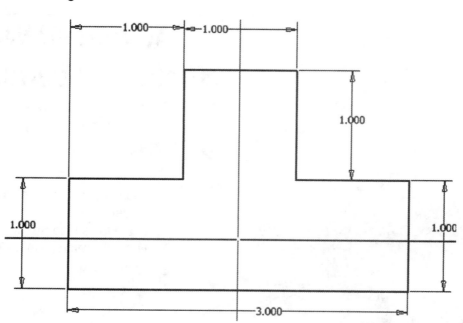

3. Exit the Sketch Panel. Extrude the sketch to a thickness of **4.00** inches as shown in Figure 2.

Figure 2

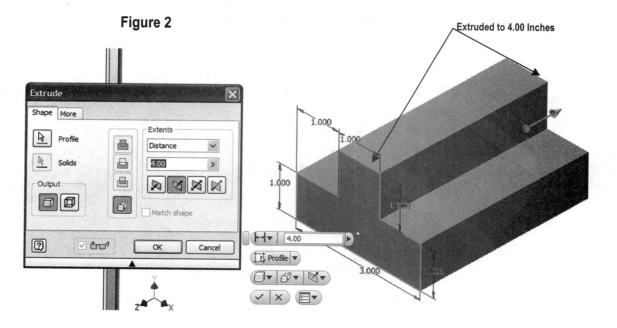

4. Your screen should look similar to Figure 3.

Figure 3

5. Use the **Rotate** command to rotate the part to gain a perpendicular view of the front surface as shown in Figure 4.

Figure 4

6. Use the **Rectangle** command to complete the sketch as shown in Figure 5. All dimensions are .188.

Figure 5

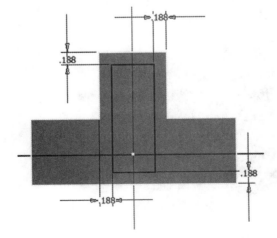

7. Extrude (cut) the rectangle a distance of **4.00** inches creating a square hole as shown in Figure 6.

Figure 6

8. Rotate the part around to gain an isometric (Home View) of the part as shown in Figure 7.

Figure 7

9. Rotate the part around to gain an isometric view of the side of the part as shown in Figure 8.

Figure 8

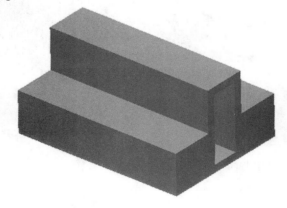

10. Begin a **New Sketch** on the front surface of the part as shown in Figure 9.

Figure 9

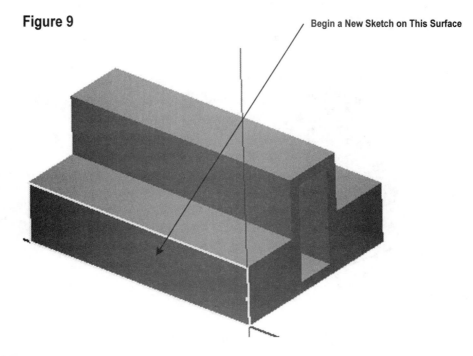

Begin a New Sketch on This Surface

Learn to create points on multiple sketches

11. Complete the sketch (2 points) as shown in Figure 10.

Figure 10

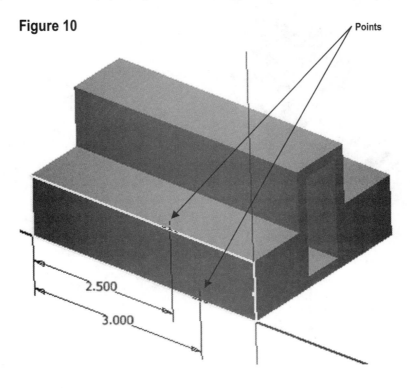

Points

2.500

3.000

12. Exit out of the Sketch as shown in Figure 11.

Figure 11

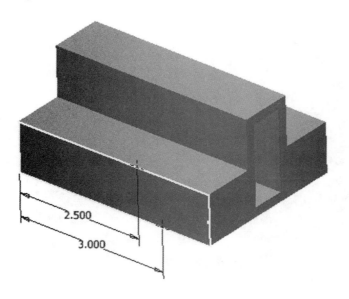

13. Rotate the part upward as shown in Figure 12.

Figure 12

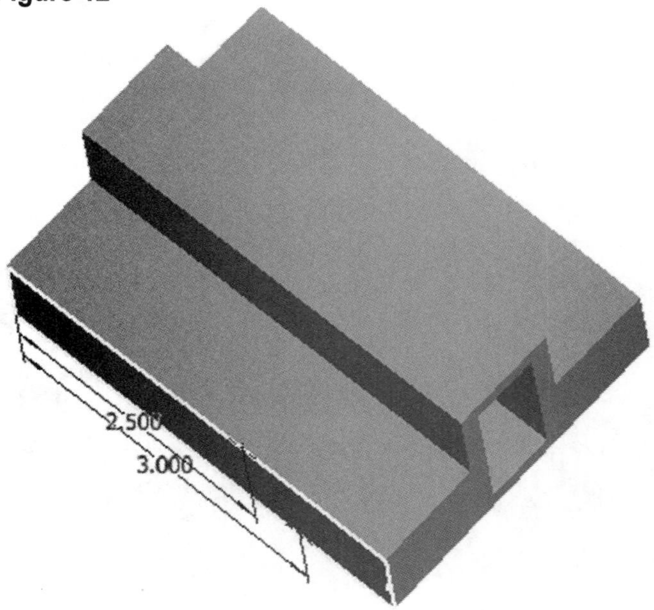

14. Begin a **New Sketch** on the surface shown in Figure 13.

Figure 13

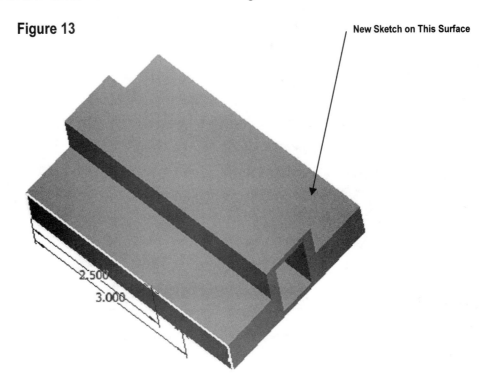

New Sketch on This Surface

15. Create a **Point** on the surface as shown in Figure 14.

Figure 14

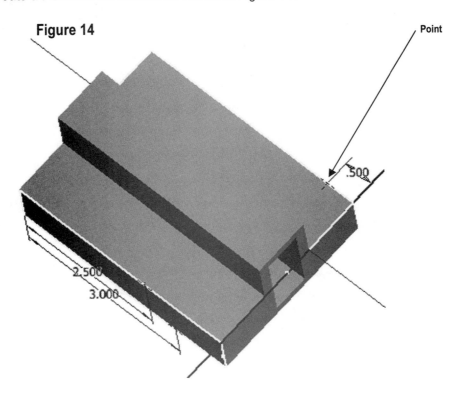

Point

16. Exit the Sketch as shown in Figure 15.

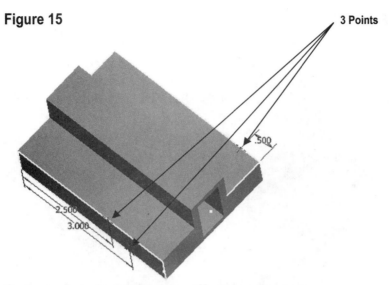

Figure 15

Learn to use these points to create an offset work plane

17. Move the cursor to the upper middle portion of the screen and left click on **Plane/Work Plane** as shown in Figure 16.

Figure 16

18. Left click on each of the 3 points as shown in Figure 17.

Figure 17

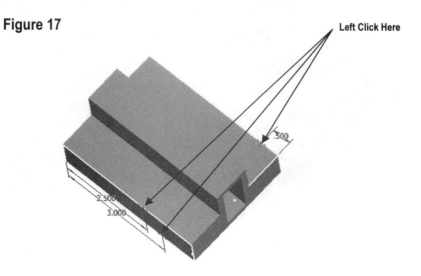

19. Inventor will create a Work Plane from the 3 points as shown in Figure 18.

Figure 18

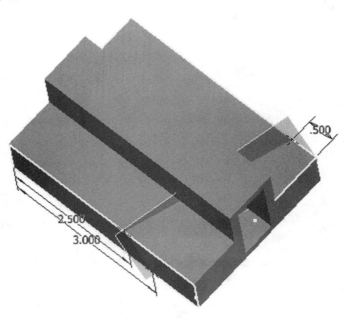

20. Move the cursor to the upper middle portion of the screen and left click on **Split** as shown in Figure 19.

Figure 19

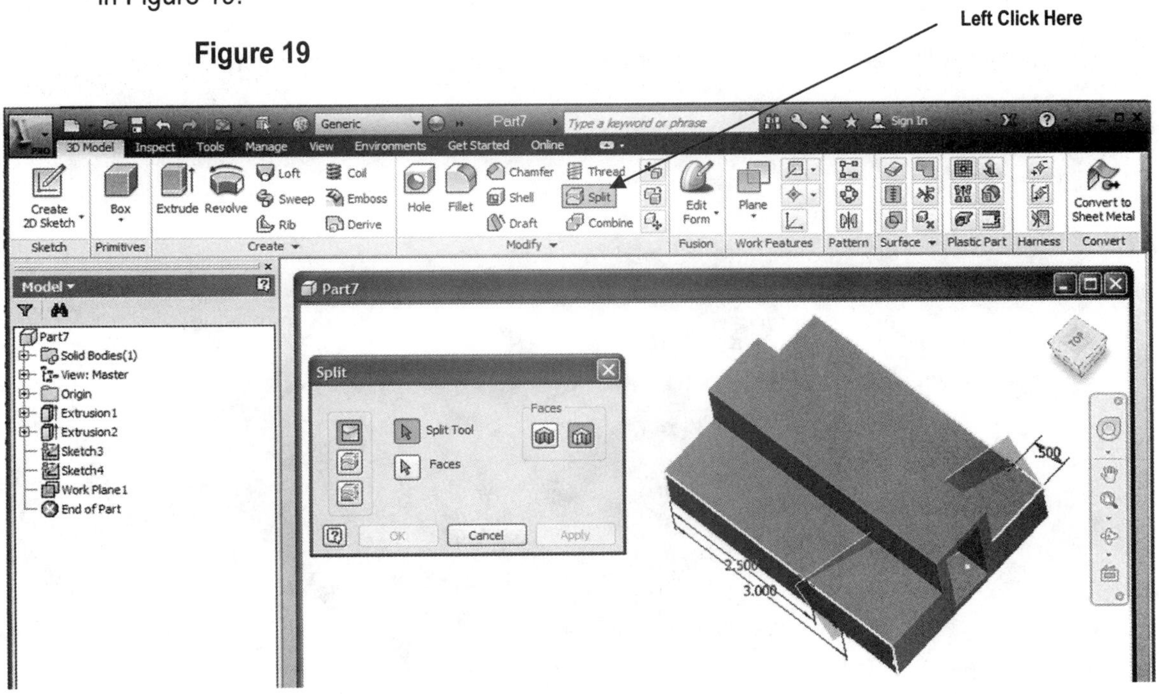

21. Move the cursor over the edge of the newly created plane causing the edges to turn red and left click as shown in Figure 20.

Figure 20

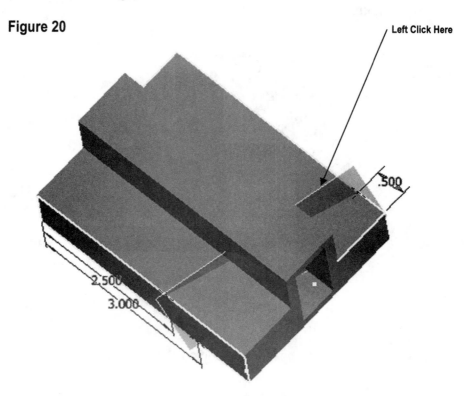

22. Inventor will create a Split plane from the work plane as shown in Figure 21.

Figure 21

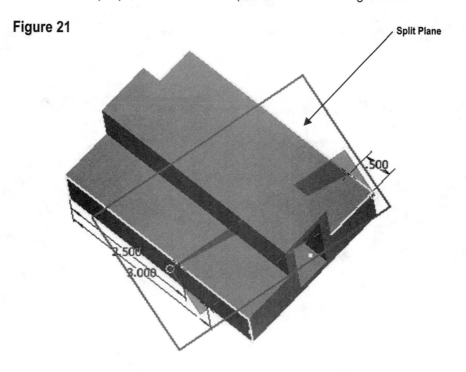

23. Left click on the **Trim Solid** icon. Left click on **OK** as shown in Figure 22.

Figure 22

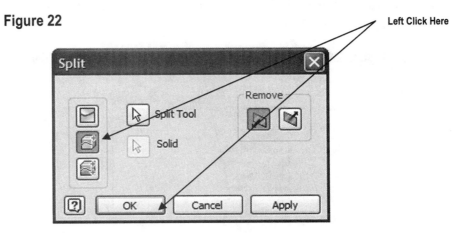

24. Your screen should look similar to Figure 23.

Figure 23

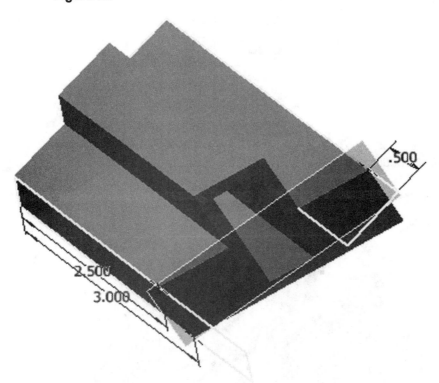

25. To clean up the appearance of the part, move the cursor over the left portion of the screen and left click, then right click on each of the sketches, and the work plane, and turn off the **Visibility** on each item as shown in Figure 24.

Figure 24

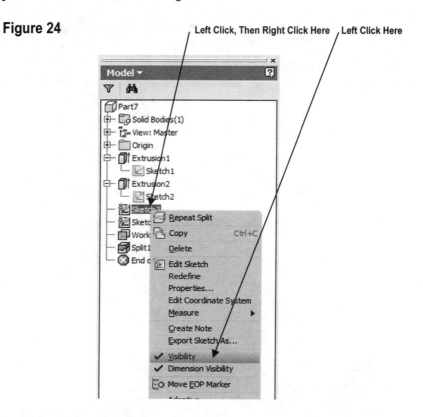

26. Your screen should look similar to Figure 25.

Figure 25

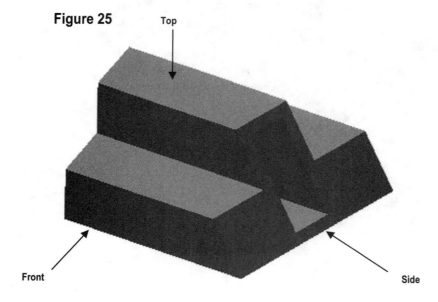

Chapter Problems

To test your Orthographic skills, create a 3 view Orthographic drawing of the part shown above (Top, Front and Right Hand side views) in either a 2 Dimensional CAD package (such as AutoCAD) or by hand on a Drafting board. Show all hidden lines in your 2 D drawing. Once you have completed your 2 Dimensional drawing, refer to Chapter 3 on how to create a 3 View drawing using Inventor, (showing all hidden lines) and then compare your "hand drawing" to the Inventor 3 view drawing.

Notes:

CHAPTER **12**

Introduction to Stress Analysis

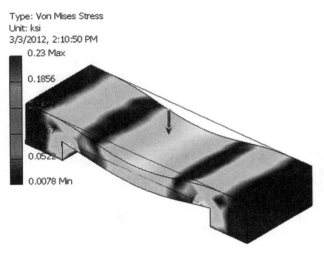

Type: Von Mises Stress
Unit: ksi
3/3/2012, 2:10:50 PM
- 0.23 Max
- 0.1856
- 0.0522
- 0.0078 Min

Objectives:

1. Learn to Create a simple part
2. Learn to run Stress Analysis
3. Learn to interpret Stress Analysis results

Chapter 12 includes instruction on how to perform a stress analysis on the part shown.

Learn to create a simple part

1. Start Inventor by referring to "Chapter 1 Getting Started".

2. After Inventor is running, begin a New Sketch.

3. Complete the sketch shown in Figure 1.

Figure 1

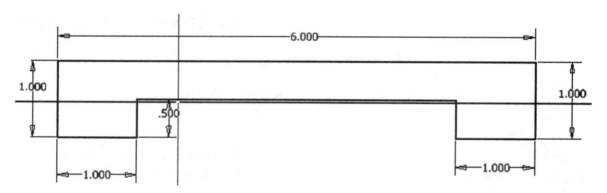

4. Exit the Sketch Panel. Extrude the sketch to a thickness of **2.00** inches as shown in Figure 2. Save the part where it can be easily retrieved later.

Figure 2

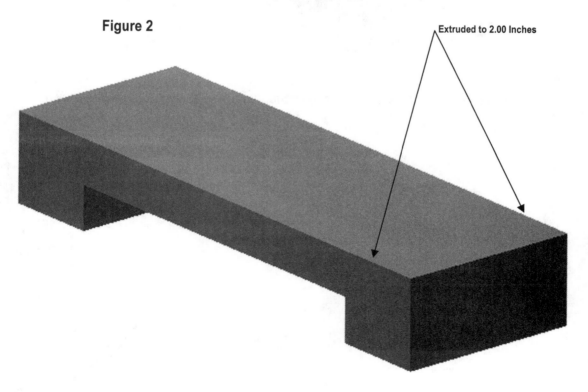

Extruded to 2.00 Inches

Learn to apply material to a simple part

5. Move the cursor to the upper left portion of the screen and left click on **Stress Analysis** as shown in Figure 3.

Figure 3

Left Click Here

6. Move the cursor to the upper left portion of the screen and left click on **Create Simulation** as shown in Figure 4.

Figure 4

Left Click Here

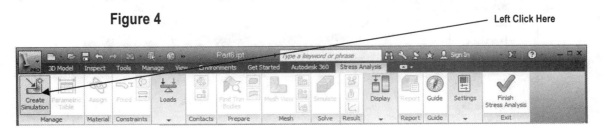

7. The Create New Simulation dialog box will appear. Left click on OK as shown in Figure 5.

Figure 5

Left Click Here

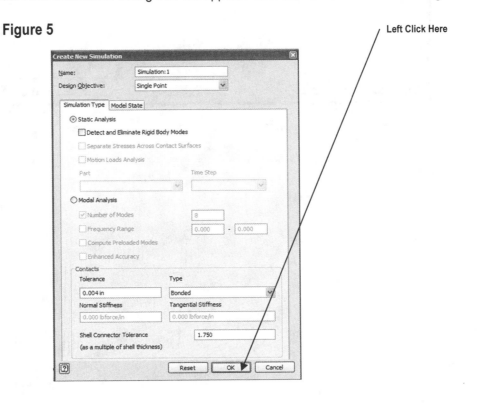

8.   Move the cursor to the upper left portion of the screen and left click on **Assign**. This "assigns" material to the solid model for analysis as shown in Figure 6.

Figure 6

Left Click Here

9.   The Assign Materials dialog box will appear. Left click on **Materials** as shown in Figure 7.

Figure 7

Left Click Here

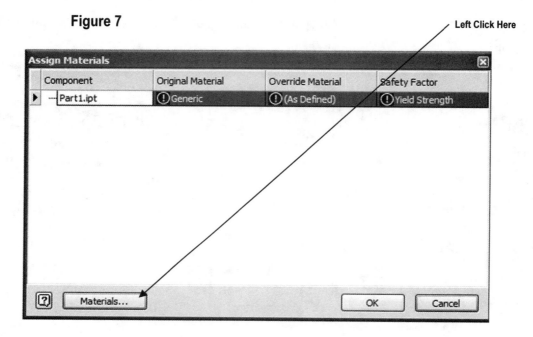

10. The Material Brower dialog box will appear. Left click on **Autodesk Material Library.** A drop down menu will appear. Left click on **Aluminum 6061** as shown in Figure 8.

Figure 8

Left Click Here

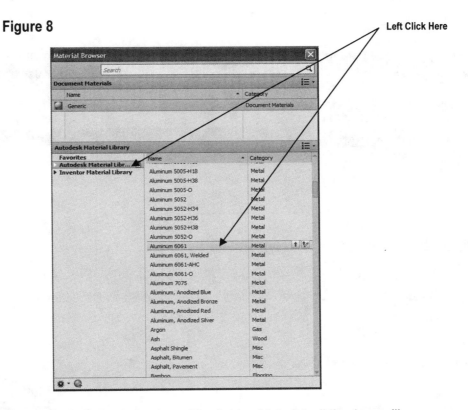

11. Close the Material Browser dialog box. The Assign Materials dialog bow will re-appear with Aluminum 6061 listed in the Original Material column. Left click on **OK** as shown in Figure 9.

Figure 9

Left Click Here

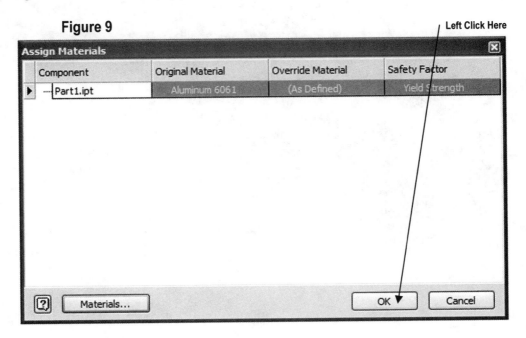

12. Move the cursor to the upper middle portion of the screen and left click on **Fixed** as shown in Figure 10.

Figure 10

Left Click Here

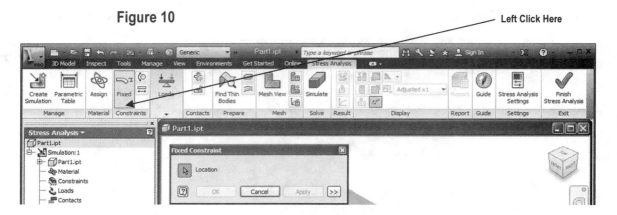

Learn to apply a fixture to a simple part

13. The Fixed Constraint dialog box will appear. Rotate the part around to gain access to the bottom side and left click on the "feet". Then, left click on **OK** as shown in Figure 11.

Figure 11

Left Click Here

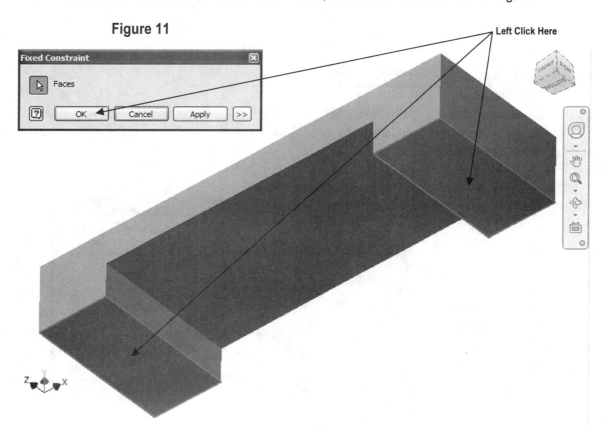

14. Move the cursor to the upper middle portion of the screen and left click on the drop down arrow located under **Loads**. A drop down menu will appear. Left click on **Force** as shown in Figure 12.

Figure 12

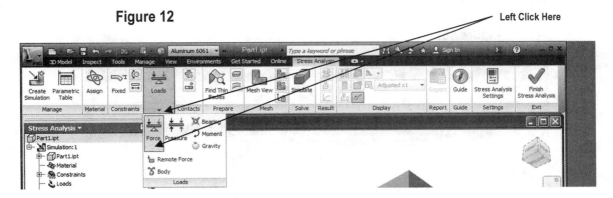

15. The Force dialog box will appear. Enter **100 lbs** for the amount of force as shown in Figure 13.

Figure 13

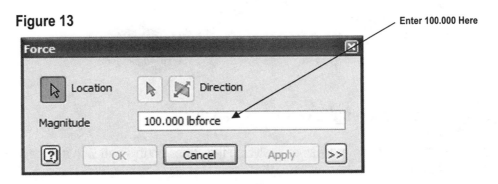

16. Rotate the part around to gain access to the top portion. Left click on the top portion. Left click on **OK** as shown in Figure 14.

Figure 14

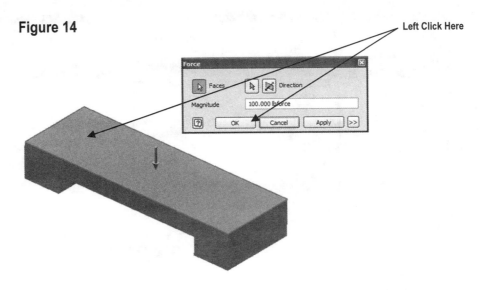

Learn to perform a stress analysis on a simple part

17. Repeat the same steps to place the Lifter1.ipt file into the assembly. Your screen should look similar to Figure 15.

Figure 15

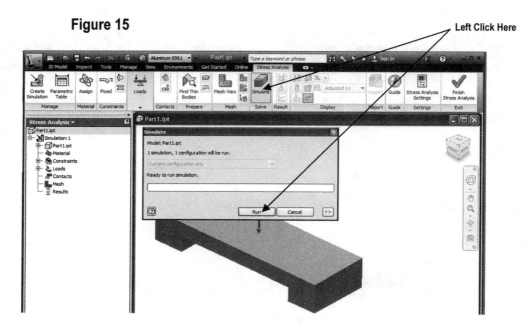

Learn to interpret results of a stress analysis

18. Inventor is in the process of analyzing the stress on this part with a force of 100 lbs as shown in Figure 16.

Figure 16

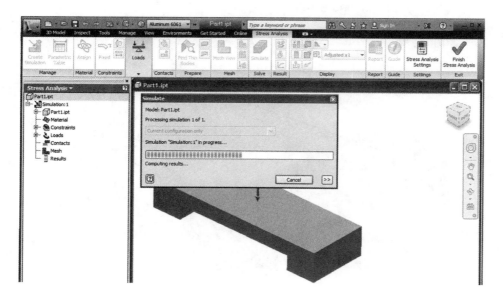

19. Inventor will display the results of the analysis. High stress areas are indentified by any reddish orange areas that appear on the model as shown in Figure 17. Rotate the part around to examine any high stress areas on the underside.

Figure 17

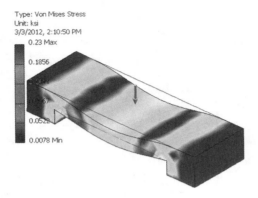

20. Move the cursor to the upper middle portion of the screen and left click on the Animate Results icon. The Animate Results dialog box will appear. Left click on "Play". Inventor will animate the stress analysis as shown in Figure 18.

Figure 18

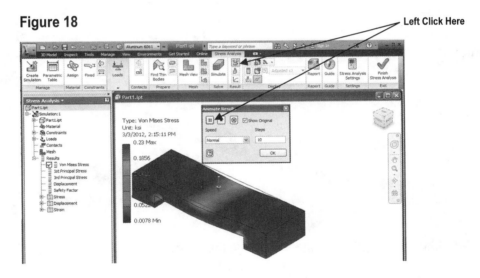

21. Results of the analysis can be viewed by selecting the Probe icon or the Convergence plot. Results can also be viewed by selecting the **Report** icon. Left click on **Finish Stress Analysis** as shown in Figure 19.

Figure 19

Chapter Problems

Problem 12-1

Complete the following sketch.

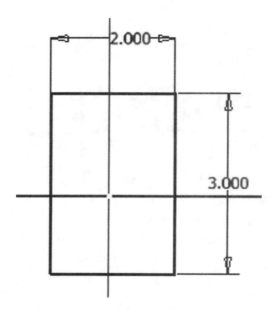

Extrude the sketch to a distance of 96 inches.

Use the Shell command to create a wall thickness of .10.

Run a stress analysis on the part using Aluminum 1100-O for material. Use the left open edge for the Fixed edge. Use a load of 500 pounds along the top. The end result will be the creation of a Cantilever as shown.

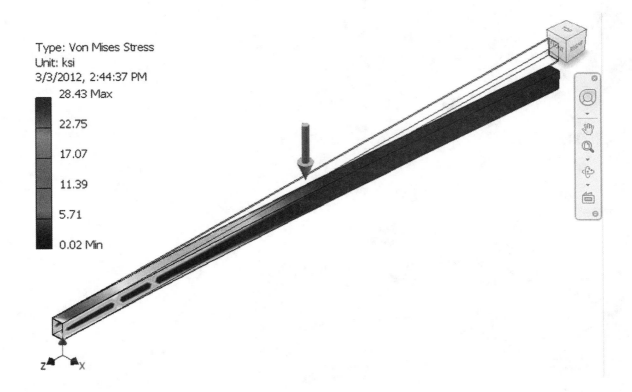

Type: Von Mises Stress
Unit: ksi
3/3/2012, 2:44:37 PM

28.43 Max

22.75

17.07

11.39

5.71

0.02 Min

Notes:

CHAPTER **13**

Introduction to the Design Accelerator

Objectives:

1. Learn to Create a Disc Cam
2. Learn to edit the Disc Cam
3. Learn to constrain the Disc Cam in an Assembly file
4. Learn to animate the Disc Cam using the Drive Constraint command

Chapter 13 includes instruction on how to design the parts shown below.
This chapter contains a brief introduction to the Design Accelerator. The Design Accelerator contains numerous predefined parts. This chapter will cover one of these parts.

1.   Start Inventor by referring to "Chapter 1 Getting Started".

2.   After Inventor is running, begin a New Sketch.

3.   Complete the sketch shown in Figure 1.

Figure 1

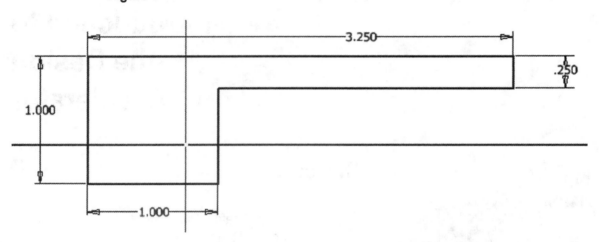

4.   Exit the Sketch Panel. Extrude the sketch to a thickness of **1.00** inch as shown in Figure 2.

Figure 2

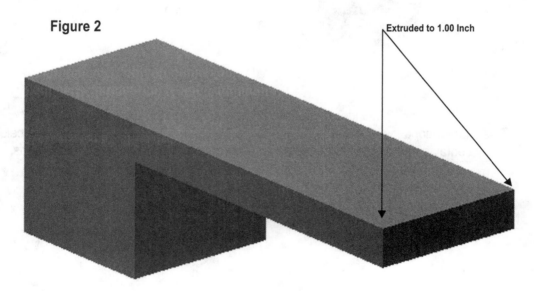

Extruded to 1.00 Inch

5. Use the **Fillet** command (.5 inch fillets) to radius the lower edge(s) as shown in Figure 3.

Figure 3

Fillet this Edge(s)

6. Use the **Rotate** command to rotate the part upward as shown in Figure 4.

Figure 4

Rotate Upward

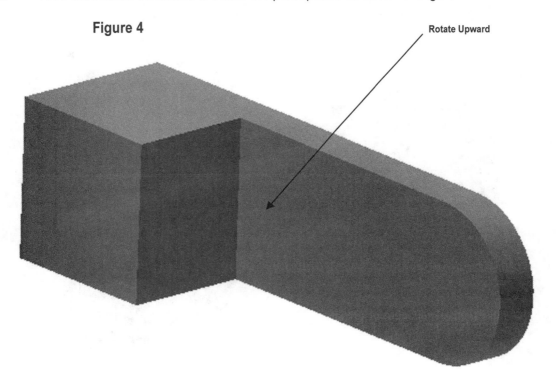

7. Complete the sketch as shown in Figure 5. The center of the circle is located on the center of the fillet radius.

Figure 5

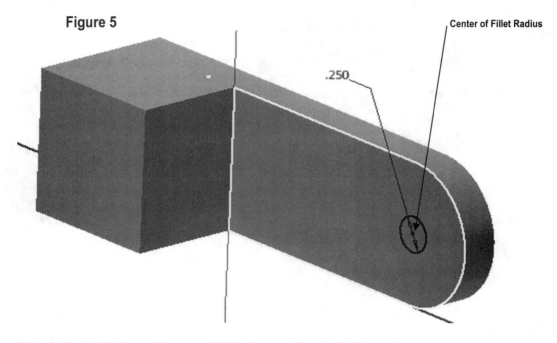

8. Extrude the circle to a distance of **.75** inches as shown in Figure 6.

Figure 6

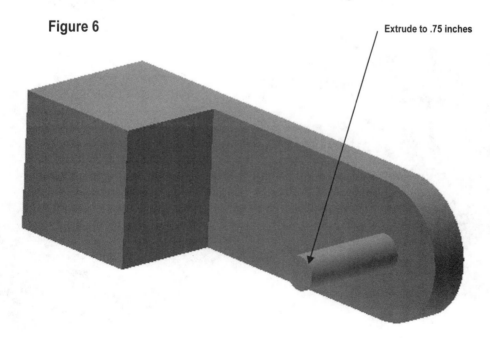

9. Complete the sketch as shown in Figure 7.

Figure 7

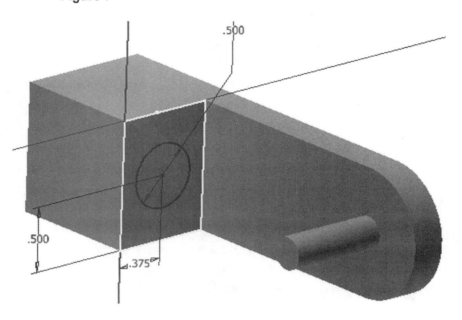

10. Use the "cut" option in the **Extrude** command to cut a hole a distance of 1.00 inch as shown in Figure 8.

Figure 8

Hole cut 1.00 Inch

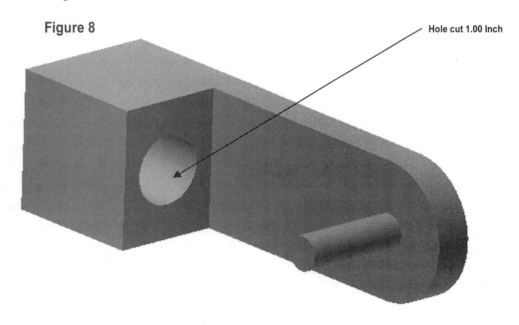

11.   Save the part as Camcase1.ipt where it can be easily retrieved later.

12.   Begin a new drawing as described in Chapter 1.

13.   Complete the sketch shown. Extrude the .500 inch diameter circle to a distance of 1 inch and the .625 inch diameter circle to a distance of .125 inches as shown in Figure 9.

Figure 9

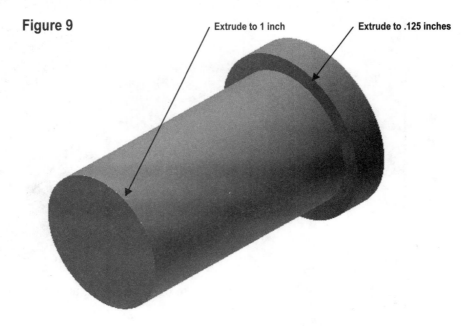

14.   The bottom of the part needs to be flat as shown in Figure 10.

Figure 10

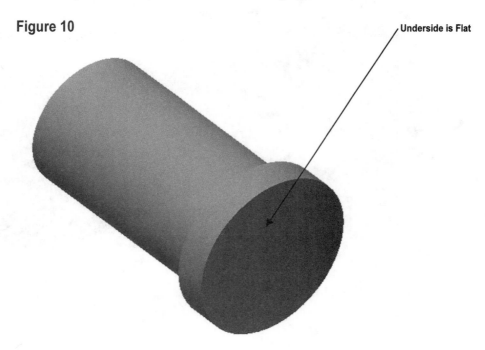

15. Save the part as Lifter1.ipt where it can be easily retrieved later.

16. Begin a new Assemble drawing as described in Chapter 7 and shown in Figure 11.

Figure 11

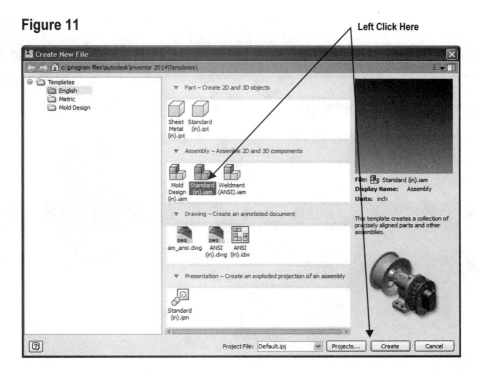

17. The Assemble Panel will appear as shown in Figure 12.

Figure 12

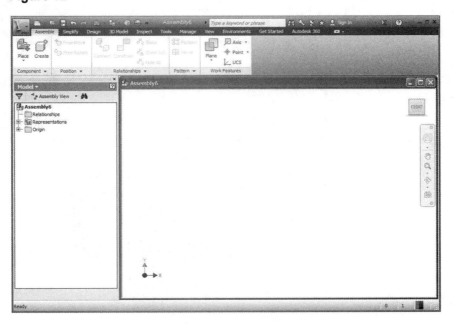

18. Use the **Place** command to place the Camcase1.ipt file into the assembly as shown in Figure 13.

Figure 13

Left Click Here

19. The Place Component dialog box will appear as shown in Figure 14.

Figure 14

Left Click Here

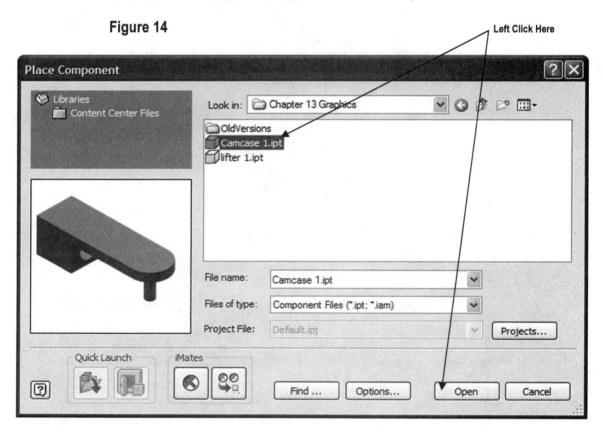

20. Repeat the same steps to place the Lifter1.ipt file into the assembly. Your screen should look similar to Figure 15.

Figure 15

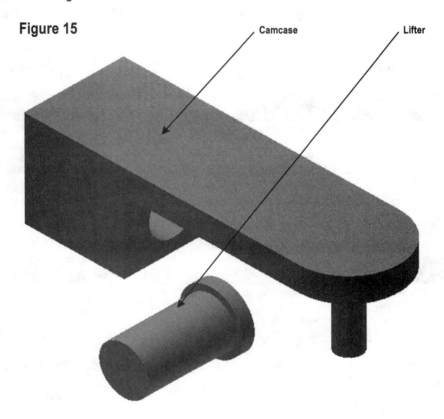

Camcase

Lifter

21. Use the **Rotate** command to rotate the case around for better access. Use the **Constraint** command to place the lifter into the lifter bore with the foot towards the bottom of the case as shown in Figure 16. Save the file before proceeding.

Figure 16

Lifter Foot

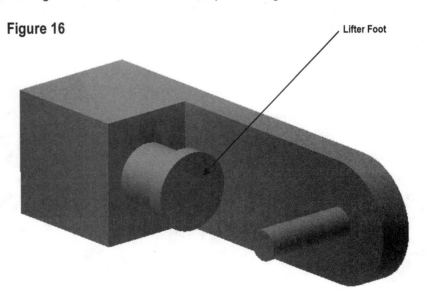

Learn to create a Disc Cam

22. Move the cursor to the upper left portion of the screen and left click on the **Design** tab as shown in Figure 17.

Figure 17

Left Click Here

23. The Design Accelerator panel will open as shown in Figure 18.

Figure 18

24. Move the cursor to the upper right portion of the screen and left click on **Disc Cam** as shown in Figure 19.

Left Click Here

Figure 19

25. The Disc Cam Component Generator dialog box will appear as shown in Figure 20.

Figure 20

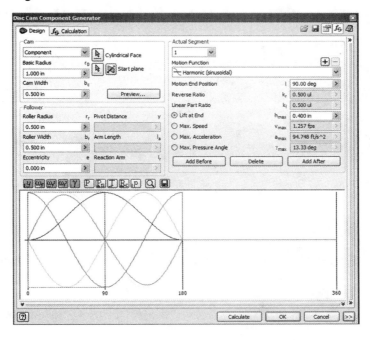

26. Highlight the text below Roller Radius and enter **2.000**. This will create a more pointed disc cam. Left click on **Calculate** and **OK** as shown in Figure 21.

Figure 21

Enter 2.000 Here

Left Click Here

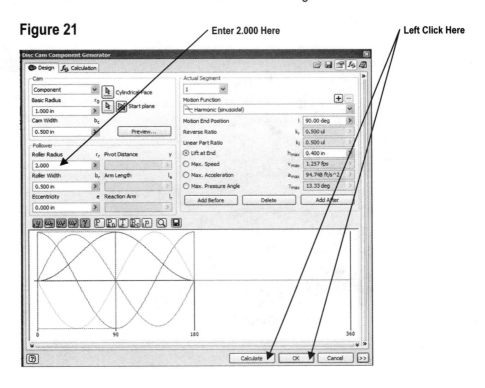

27. The disc cam will be attached to the cursor. If the "File Naming" dialog box (not shown) appears, left click on OK. Left click as shown in Figure 22.

Figure 22

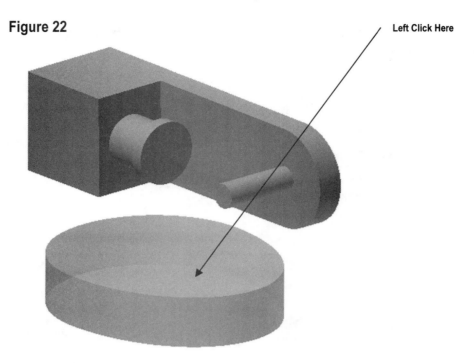

Left Click Here

28. Your screen should look similar to Figure 23.

Figure 23

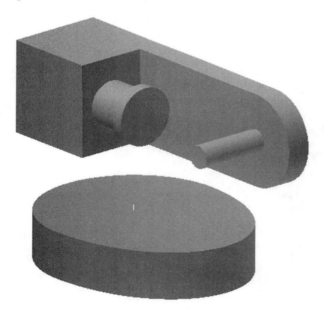

Learn to edit a Disc Cam

29. The disc cam will need the addition of a center hole and key slot. Move the cursor of the upper face of the disc cam causing the edges to turn red and double click once as shown in Figure 24.

Figure 24

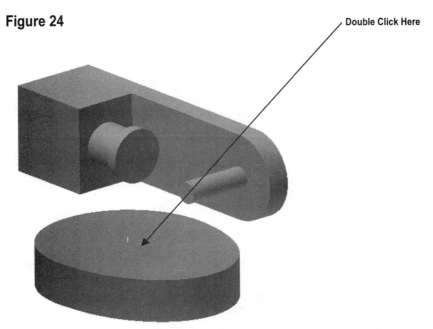

Double Click Here

30. The rest of the parts in the assembly will become inactive as shown in Figure 25.

Figure 25

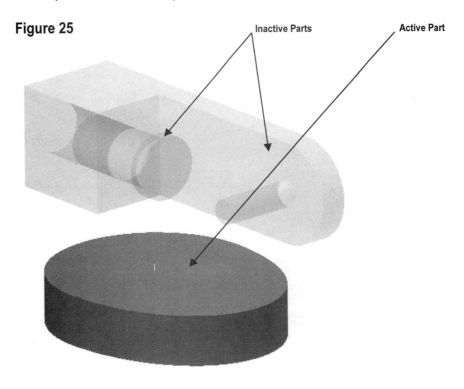

Inactive Parts   Active Part

31. Move the cursor to the upper face of the disc cam causing the edges to turn red and right click once. A pop up menu will appear. Left click on **Edit** as shown in Figure 26.

Figure 26

Left Click Here

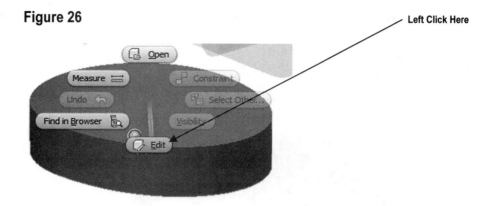

32. Move the cursor to the upper face of the cam disc causing the edges to turn red and right click once. A pop up menu will appear. Left click on **New Sketch** as shown in Figure 27.

Figure 27

Left Click Here

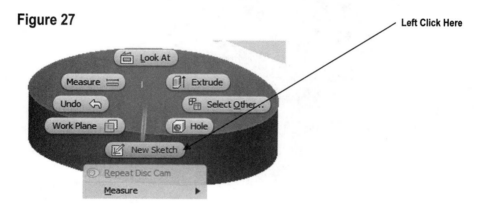

33. Inventor will return to the Sketch Panel as shown in Figure 28.

Figure 28

Sketch Lines

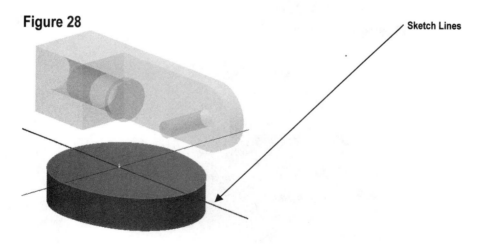

34. Use the **Look At** command to gain a perpendicular view of the upper face of the cam disc as shown in Figure 29.

Figure 29

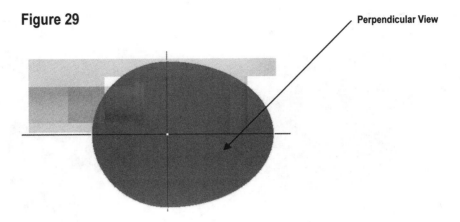

35. Complete the sketch as shown in Figure 30.

Figure 30

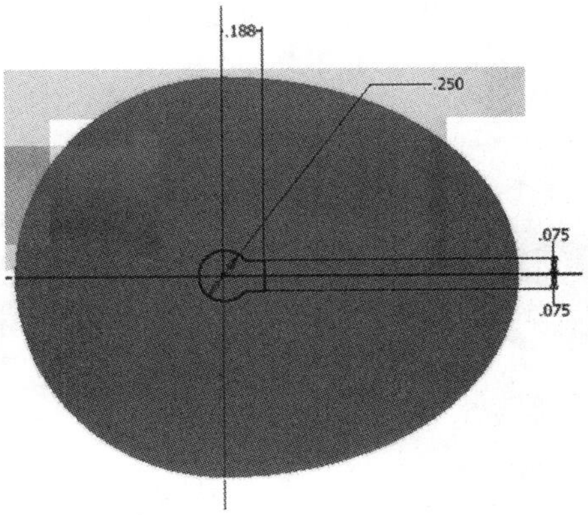

36. Once the sketch is complete, exit out of the Sketch Panel as shown in Figure 31.

Figure 31

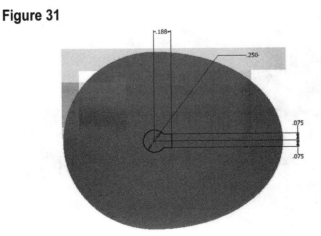

37. Use the **Rotate** command to rotate the part around as shown in Figure 32. Use the **Extrude** command to cut a hole and key slot in the part as shown.

Figure 32

Extrude Hole and Key Slot

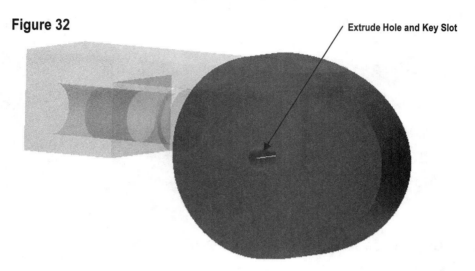

38. Move the cursor to the upper left portion of the screen and left click on the **Manage** tab. Left click on **Return** twice as shown in Figure 33.

Figure 33   Left Click Here   Left Click Here Twice

39. Your screen should look similar to Figure 34.

Figure 34

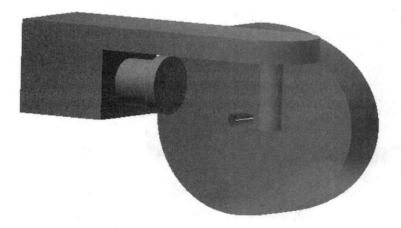

40. Move the cursor to the upper left portion of the screen and left click on the **Assemble** tab. Inventor will return to the Assemble Panel if not already in the Assemble Panel.

Figure 35

Left Click Here

41. Use the **Constraint** command to constrain the center of the disc cam to the center of the shaft as shown in Figure 36.

Figure 36

42. Rotate the assembly around to gain access to the side of the disc cam. Use the **Constraint** command to constrain the disc cam an offset distance of **.125** inches from the inside of the cam case as shown in Figure 37.

Figure 37

.125 Inch Offset

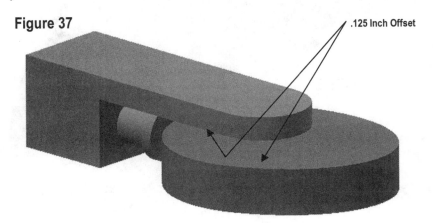

43.   Use the **Rotate** command to rotate the part as shown. The disc cam's location on the shaft should be similar to Figure 38.

Figure 38

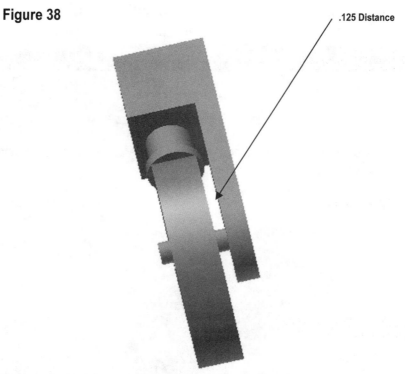

.125 Distance

44.   Use the **Rotate** command to rotate the assembly around as shown in Figure 39.

Figure 39

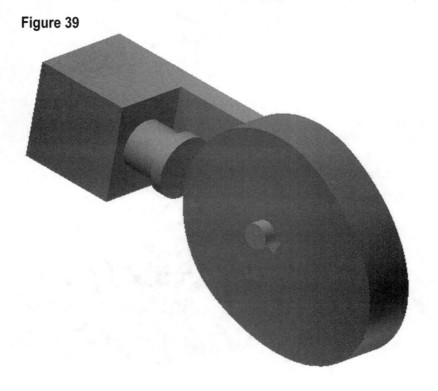

45. Move the cursor to the upper left portion of the screen and left click on **Constrain** as shown in Figure 40.

Figure 40

46. The Place Constraint dialog box will appear. Left click on the **Tangent** icon. Left click on the "Outside" solution as shown in Figure 41.

Figure 41

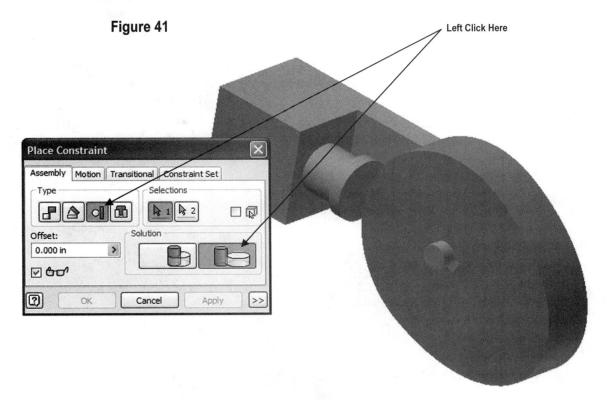

47.   Left click on the lifter foot as shown in Figure 42.

Figure 42

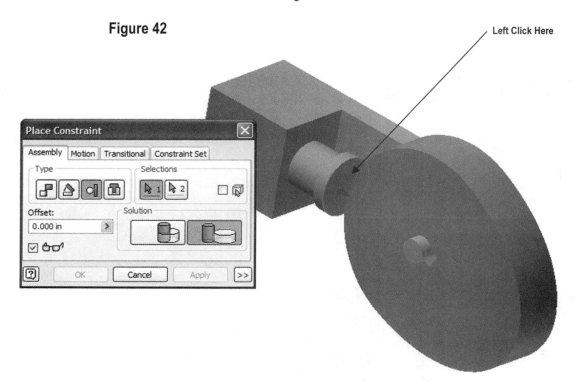

48.   Left click on the surface of the disc cam as shown in Figure 43.

Figure 43

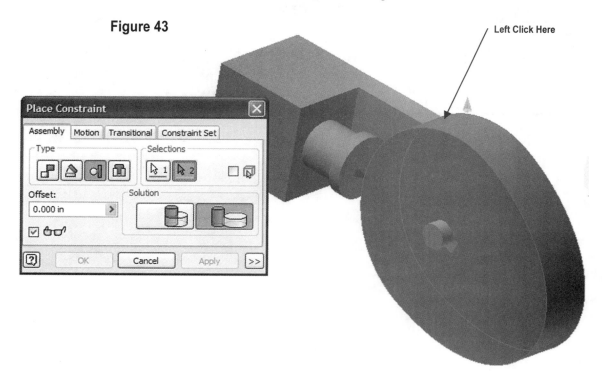

49. Left click on **OK** as shown in Figure 44.

Figure 44

Left Click Here

50. Use the cursor to rotate the disc cam upward as shown in Figure 45.

Figure 45

51. Move the cursor to the upper left portion of the screen and left click on **Constrain** as shown in Figure 46.

Figure 46

Left Click Here

52. The Place Constraint dialog box will appear. Left click on the **Angle** icon. Left click on the "Directed Angle" solution as shown in Figure 47.

Figure 47

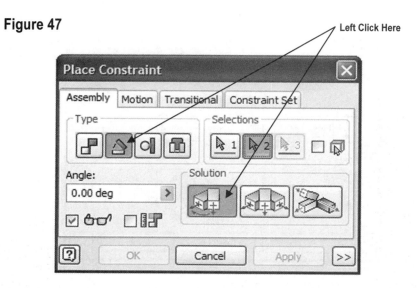

53. Use the **Zoom** command to zoom in on the key slot. Left click on the back (inside) of the key slot as shown in Figure 48.

Figure 48

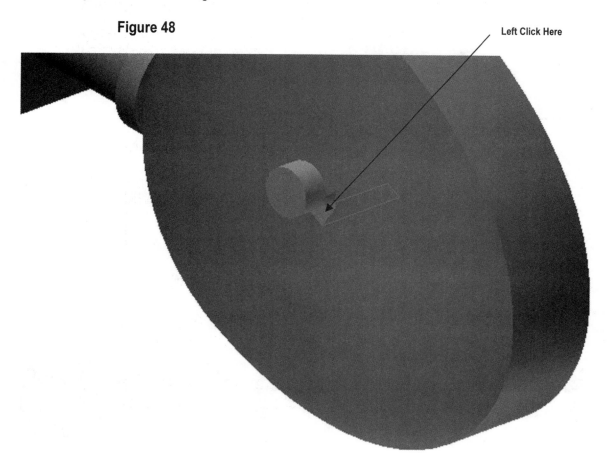

Learn to animate the assembly

54. Left click on the underside of the cam case as shown in Figure 49.

Figure 49

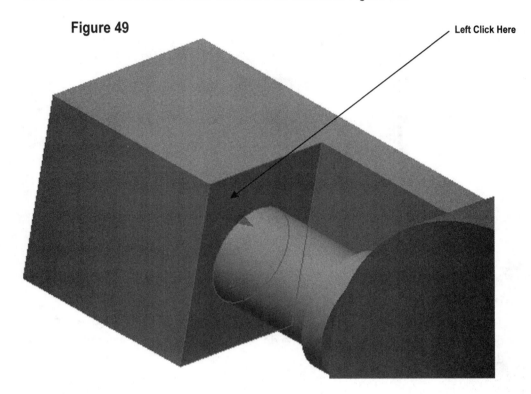

Left Click Here

55. Left click on **OK** as shown in Figure 50.

Figure 50

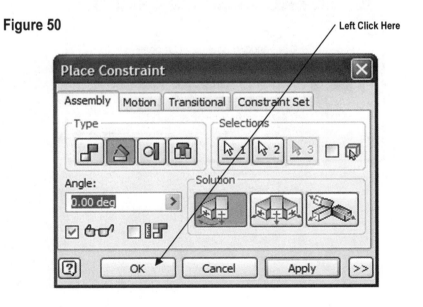

Left Click Here

56. Refer back to Chapter 7 on Driving Constraints if needed. Move the cursor to the lower left portion of the screen to the part tree. Scroll down to **Angle:1** and right click once. A pop up menu will appear. Left click on **Drive Constraint** as shown in Figure 51.

Figure 51

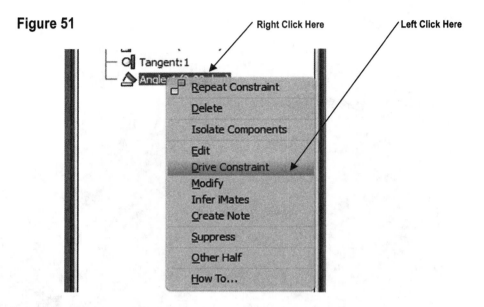

57. The Drive Constraint dialog box will appear. Enter **0** degrees under "Start". Enter **360000** degrees under "End". Left click on the double arrows at the far lower right corner of the dialog box as shown in Figure 52.

Figure 52

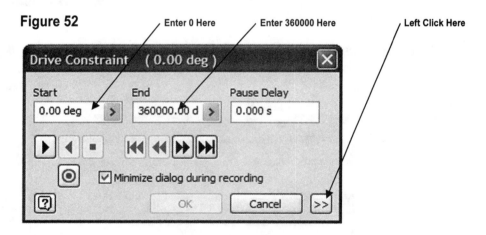

58. The Drive Constraint dialog box will expand providing more options. Enter **10** for the number of degrees as shown in Figure 53.

Figure 53

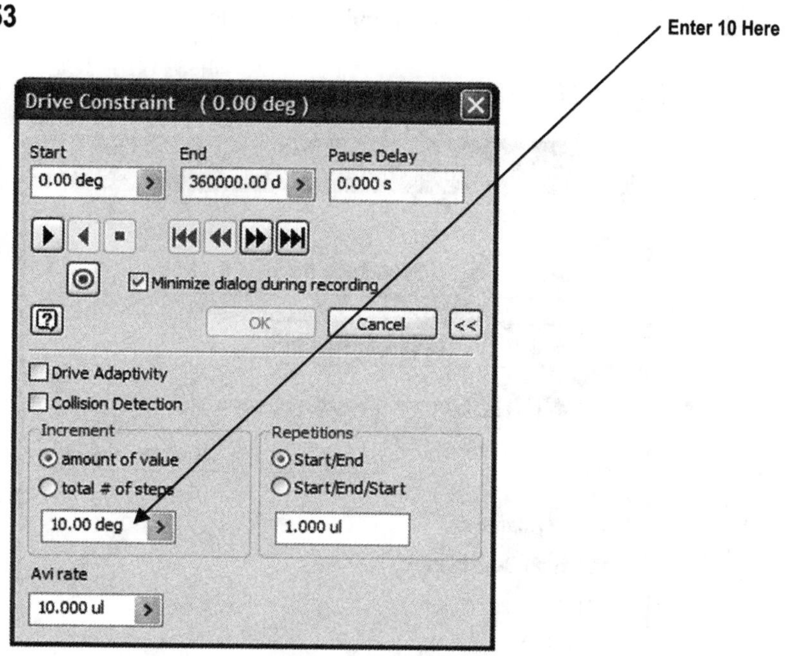

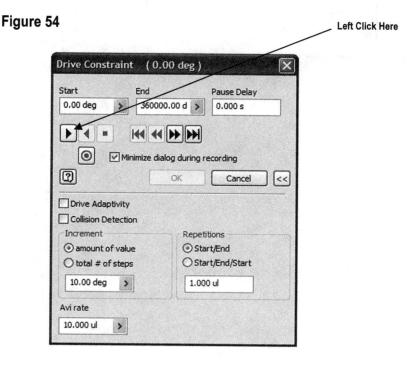

59. Use the **Zoom** option to zoom out. Use the **Pan** option to move the assembly off to the side. Left click on the "Play" icon as shown in Figure 54.

Figure 54

60. Inventor will animate the assembly causing the disc cam to rotate.

61. Left click on the "Stop" icon. The animation will stop. Left click on the "Minimize" icon. The Drive Constraint dialog box will get smaller. Left click on the "Rewind" icon. This will rewind the animation back to 0 degrees as shown in Figure 55. Refer to Chapter 7 for instructions on how to create an .avi or .wmv file.

Figure 55

Stop Icon   Rewind Icon   Minimize Icon

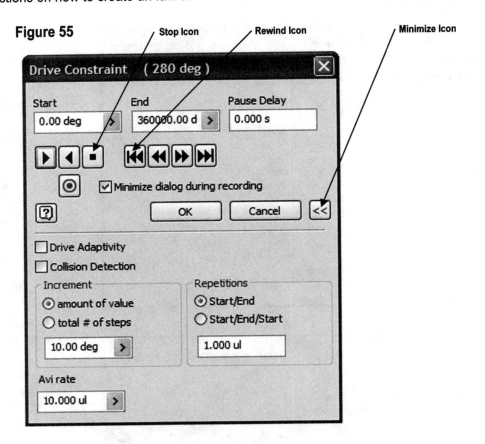